Darwin's Philosophical Legacy

Darwin's Philosophical Legacy

The Good and the Not-So-Good

Dr. Gerard M. Verschuuren

LEXINGTON BOOKS
Lanham • Boulder • New York • Toronto • Plymouth, UK

Published by Lexington Books
A wholly owned subsidiary of The Rowman & Littlefield Publishing Group, Inc.
4501 Forbes Boulevard, Suite 200, Lanham, Maryland 20706
www.rowman.com

10 Thornbury Road, Plymouth PL6 7PP, United Kingdom

British Library Cataloguing in Publication Information Available

Library of Congress Cataloging-in-Publication Data

Verschuuren, G. M. N. (Geert M. N.)
Darwin's philosophical legacy : the good and the not-so-good / Gerard M. Verschuuren.
p. cm.
Includes bibliographical references and index.
ISBN 978-0-7391-7520-0 (cloth : alk. paper) -- ISBN 978-0-7391-7521-7 (electronic)
1. Evolution. 2. Evolution (Biology) 3. Darwin, Charles, 1809-1882. I. Title.
B818.V47 2012
576.8'2092--dc23
2012007556

The paper used in this publication meets the minimum requirements of American National Standard for Information Sciences Permanence of Paper for Printed Library Materials, ANSI/NISO Z39.48-1992.

Printed in the United States of America

Dedicated to the memory of one of my first professors,

Dr. John Huizinga, M.D.,

who initiated me into the Darwinian paradigm

Contents

Preface

Charles Darwin's legacy is tremendous; his thoughts and ideas set the study of evolution into real motion. He has been called the Newton or even the Einstein of biology. That doesn't mean we have to worship at his "altar," but we cannot and should not ignore his impact on our culture and our scientific thinking, for our world has undoubtedly changed through the arrival of his 1859 book entitled *On the Origin of Species by Means of Natural Selection*. The roots of all our evolutionary thinking go back to Darwin. His evolutionary ideas turned out to be very revolutionary.

There is hardly any university, college, or even high school left where they do not teach Darwinism—and rightly so. Yet, most of these places do more preaching than teaching. They teach more than they should, and at the same time, they teach less than they should. As to my own position: I am a staunch defender of what they currently call neo-Darwinism or "synthetic evolutionary theory," and yet I also realize that neo-Darwinism is an enterprise that has several built-in limitations, especially so when seen from a philosophical perspective. If we lose sight of either side, we won't do justice to Darwin's legacy.

This book is not so much about the biological merits or shortcomings of Darwinism; it is not a biological compendium of neo-Darwinism, for many others have written about that part. It is more about the strong and the weak points of Darwin's legacy in a philosophical sense. It leaves Darwinism intact, but looks at it from very unusual angles. It gives Darwinism its due, but removes the "sting" that many of its opponents dislike.

In what I like to call "The Good" parts of his legacy, I will explore what Darwin's great contributions are to the study of evolution and to the theory of evolution. At the same time, I will also delve into the areas where his thoughts were not so perfect or even wrong, especially in a philosophical sense—which I call "The Not-So-Good" parts of his legacy. There are definitely two sides to Darwin's legacy and we need to carefully balance between both.

Many books have been written about Darwin and his "revolution," but they are usually entirely for or against his work. This book, in contrast, wants to evaluate Darwin's legacy from a more philosophical viewpoint by analyzing the strong points as well as the weak points of what came to be known as neo-Darwinism. It invites the reader to reflect more in particular on its philosophical implications. It will change the way most readers think about this fundamental issue. In doing so, it opens up new vistas that few have ever seen before. I invite you to follow me on this exciting journey.

Acknowledgments

I wish to express special gratitude to those who steered and corrected me in the long process of preparation for this book: Francisco Ayala of the University of California at Irvine, Peter Kreeft of Boston College, Kenneth R. Miller of Brown University, Wim J. Van Der Steen of Free University in Amsterdam, and the late Marius Jeuken and Cornelius Van Peursen of Leiden University in the Netherlands. Obviously, they are not responsible for the outcome; if I erred, it's entirely my doing. They and many others make me realize that originality only consists in the capacity of forgetting about your sources.

And—as with all my books—I could not have written this book without the unwavering and loving support of my wife, Trudy. I also want to specifically mention the great and helpful comments I received from our daughter Elizabeth Frangiosa, Attorney at Law.

Chapter One

Selection Filter: The Good Legacy

In 1859, when Charles Darwin (1809–1882) published his best-known book—*On the Origin of Species by Means of Natural Selection, or the Preservation of Favoured Races in the Struggle for Life*[1]—he put forward his solution to one of the burning scientific questions of the day: What are the causal origins of the world's organisms, past and present? How did organisms come to be as they are now? Not only did he argue that all organisms descended from previous generations by gradual modifications—often called *common descent with modification*—but he also suggested a causal mechanism for this process by embracing the concept of *natural selection.* Curiously enough, in his book, he did not use the word "evolution," but always referred to it as "descent with modification."

The idea that all species have common ancestors, or common descent, was not really new. Several Greek philosophers had already adopted this idea, but the hypothesis got a real boost with William Harvey's claim (1651) that all life emerges from life ("*omne vivum ex ovo* [*or vivo*]"[2]), so there wouldn't be any life without "parents." From then on, evolutionary ideas were set out by a few natural philosophers including Pierre Maupertuis in 1745, Erasmus Darwin in 1796, and Jean-Baptiste Lamarck in 1800. However, it was Charles Darwin who introduced the idea of natural selection as a mechanism of evolution.[3] He was still developing his theory in 1858 when Alfred Russell Wallace sent him a similar theory, so he rushed to publish his book in 1859.

Ever since, the idea of evolution has "evolved" from a hypothesis to a theory to a fact,[4] but the science of the way evolution operates is still under construction. Let us assume from now on that you do accept evolution as an established fact, that you do accept that there is a more or less gradual change, through successive living forms, from the origin to the present, with species giving rise to new species. Once we accept this idea or fact, we have to look "under the hood" of evolution and come up with a mechanism that explains how exactly species can multiply and change over time; without such an explanation, evolution becomes questionable.

Many scientists have tried to come up with an explanation, but the winning horse in the race seems to have come from Charles Darwin, as he is the one who hypothesized the mechanism of natural selection (although there is more to it). Darwin's theory of natural selection basically attempts to explain nature's beautiful design in terms of *physical* causes and *natural* laws—without invoking Designer interventions. Let me explain this part first.

Unlike the other natural sciences, the life sciences study phenomena not only in relation to their (physical) causes, but also in relation to their "effects" in surviving. What does this mean? The fact, for instance, that the caterpillars of a white cabbage butterfly are green rather than white (which is caused by certain genes), makes these slow organisms feeding on green cabbage less conspicuous to predators and thus more successful in survival and reproduction. In cases like these, we are not so much interested in the *causes* of camouflage (such as genes) as we are in its *effects* on survival.

Biological features can be understood in terms of their effects—that is, in terms of survival problems that need to be effectively solved. In other words, they serve a *function*; the green color of a caterpillar has a function, namely, to deceive potential predators; such is their end or goal—or in more neutral, biological terms, such is their function. You could also say that camouflage is "for" deceiving, just like a knife is "for" cutting. But make no mistake, green caterpillars achieve a goal without having that goal as a "purpose in mind"; they were just born that way. So what they have are functions, not intentions. Intentions and purposes may be something in the mind of a human product-maker, but functions and designs are a feature of the product itself—actually a "vital" feature relating to survival.

How did such a design ever come along? Was it designed by a divine Designer? Darwin did not really go into this discussion, but instead looked for a causal, physical, natural explanation based on past history. In his scientific quest for causality, in his search for "laws of nature" and physical, natural explanations, he came up with the mechanism of natural selection. Natural selection promotes good designs more so than bad designs, which makes them increase their frequency in future generations. Hence the reasoning goes basically as follows: Their green color causes camouflage; camouflage is a successful design; therefore natural selection causes this functional causality to spread. In other words, the green color of caterpillars has a selective advantage over other colors, and therefore increases its frequency through better chances of reproduction.

To put it briefly, natural selection is a causal, physical mechanism based on nature's functionality and design; a fancier word for this kind of functionality is *teleology*. Natural selection favors biological designs that are functional by solving a problem posed by the environment. The idea of this natural selection mechanism is based on the fact that organisms are different with regards to their chances of survival and reproduction. The more an organism is adapted to its environment—making for a better design fit to solve a problem posed by the environment—the more likely this organism is to contribute to the genetic constitution of the next generations.

Don't think this is an explanation of past causes in terms of future goals; if those (future) goals are successful, then natural selection promotes its (past) causes. That seems to be a law of nature in Darwin's view. Somehow, natural selection promotes the better designs by "weighing" the benefits against the costs like an engineer, economist, or architect would. Of course, organisms do not really calculate costs and benefits, but natural selection makes them act *as if* they do make strategic decisions. The wing of a bird follows the same laws of aerodynamics as the wing of a plane.

Again, the quest for an explanation as to what makes a population of caterpillars predominantly green is actually a quest for *causes*—and Darwin found his answer in the law of natural selection. Natural selection works on (genetic and hereditary[5]) variation and diversity, and then promotes the designs that are more "successful," more "adapted," more "effective," more "goal-directed," more "teleological" than others. So what he basically did is he replaced explanation by divine, supernatural intervention with a physical, natural explanation based on natural causes, since Darwin considered it to be his scientific duty to come up with physical and natural causes and laws. In his view, the phenomenon of design in nature, also called teleology, is not to be explained in anthropomorphic or deistic terms, but in terms of a "physical" process, natural selection. Thus he rephrased teleology from an "a priori drive" to an "a posteriori result."

As a result, Darwinism sees a special relationship between causality and functionality. It goes like this. There is some genetic program which causes a certain effect (also called *proximate* causality); this effect may have a certain survival value (functionality); and as a consequence, each particular program is constantly adjusted by the survival value of the achieved endpoint and is therefore the result of natural selection (also called *ultimate* causality).

The aim of the theory of natural selection is to explain any trait in a population in terms of *adaptation*. Evolution is supposed to produce the best adapted phenotypes. If a trait X has a function Y, then Y is called "adaptive"—which means that it promotes survival and reproduction. And therefore X will become an adaptation, for any trait having a function increases the likelihood that organisms with that trait will survive longer and/or reproduce more in comparison with organisms with other traits.[6] Consequently, a trait is not an adaptation in itself; it is rather an adaptation in comparison with other traits in a particular environment and with respect to particular criteria. Functions are usually measured in terms of *fitness*—being the expected genetic contribution to future generations. The ultimate function of a trait is to enable organisms to adapt to their environment, to increase the chances of individual survival and reproduction—briefly, to enhance fitness.

Given the fact that natural selection is a process of selecting better and better solutions, if available, for the problems a population encounters in its environment, one can construct an "optimal design" which can be set alongside the actual solutions that a certain population has developed. This is called an "optimality model." Such a model takes the natural world *as if* it were designed by an engineer or economist who is concerned to get the maximum output for the minimum input. An optimal foraging model, for example, tries to show that particular foraging patterns maximize or optimize the net calorie intake of organisms; it assumes that organisms will forage first for food items that give the greater harvest per unit of time.

In fact, organisms do not calculate costs and benefits, but they act *as if* they make strategic decisions. Thus, optimality modeling is a way of finding out whether populations solve a given environmental problem in the way the optimality model predicts. When there is a discrepancy between optimal and actual solutions, there are plenty of reasons why the optimal solution may have been compromised—for instance, genetic and developmental constraints, past history, opposing selection pressures (sexual selection versus predator selection), and many others.

In spite of its testability, the concept of natural selection has been attacked from several sides. One of the more unsophisticated attacks came from the Anglican Bishop Hugh Montefiore (1920–2005).[7] He condemned the notion of natural selection based on camouflage as spurious. His reasoning was that natural selection couldn't have favored the successful design of a white camouflage for polar bears, because bears don't have predators. He forgot, however, that camouflage works in two directions—for predators as well as prey. It would be hard for a polar bear with a dark fur to sneak up to a seal.

Another attack claims that Darwin replaced God as a selecting agent with something like "the goddess Nature." Some say he just substituted a divine watchmaker with a "blind watchmaker" (à la Richard Dawkins[8]). Darwin must have worried about this issue, for he always felt uneasy about his term "natural selection." As a matter of fact, this term leads us almost automatically to the obvious question "Selection by whom or by what?" Darwin certainly tried to avoid this implication by saying he had as much right to use metaphorical language as physicists do. In his own words, "who objects to an author speaking of the attraction of gravity as ruling the movements of the planets? Everyone knows what is meant and is implied by such metaphorical expressions."[9] Yet, his colleague Alfred Wallace convinced Darwin to replace the term "natural selection" with Spencer's idea of "survival of the fittest" in the fifth edition of his book. Darwin also liked the term "natural preservation."

A more serious attack is the accusation that the idea of natural selection amounts to a tautology—that is, a statement which is true in every possible interpretation. The argument goes like this: "Who survive? The fittest! Who are the fittest? Those who survive!" I don't think we need to take this attack serious, for it is based on terminological confusion. Biological fitness has actually a double meaning: It refers to the role of an organism as the subject (or cause) of reproduction, but also to the role of an organism as the object (or effect) of reproduction. This ambiguity also affects the fitness concept: Darwin's concept of fitness—let's call it D-fitness—is *potential* reproductive success (a cause), but there is also the concept of fitness in the sense of *actual* reproductive success (an effect), which is the way the statistician Robert Fisher and some population geneticists use it—let's coin it F-fitness. Apparently, the slogan "survival of the fittest" is not a tautology if we take fitness as D-fitness, or reproductive capacity. The principle of natural selection (or survival of the fittest) asserts that those organisms that are potentially successful in reproduction are more likely to be also actually successful in reproduction. Having a good design does matter in evolution.

As to the question whether natural selection is the only factor operational in evolution, the answer is plainly no. It may be a main factor, but this is not to say that there are no other additional factors involved, such as mutation, migration, isolation, sampling, catastrophes, and the like. What matters the most is that Darwin's theory opened the gate to a very successful research program. Because of this program, Bernard Kettlewell began to study how the composition of the peppered moth population experienced a change when industrial areas in England became more polluted; H. W. Bates started the study of mimicry, the phenomenon that an unprotected species (say of flies) takes advantage of its partial resemblance to a protected species (say of wasps), and then discovered that when the protected species varied geographically, its mimicking satellites had undergone exactly the same changes as their unpalatable models; and then there was Georges Teissier who designed simple population cages to experiment with different kinds of selection pressure on fruit flies; and later on, we see population geneticists using gene-pool models that they subjected to specific selection pressures. Darwin's research program was so powerful that many biologists nowadays are still successfully working within its framework.

I am not saying here that Darwinism has the last word for all our evolutionary problems. Science is never final; it is fundamentally of a tentative nature, being engaged in a never-ending spiral of searching for the truth and constantly revising facts and theories. Nevertheless, it is a timeless temptation for scientists, including Darwinists, to think that they have finally been able to reduce the unknown to nothing, or at least almost nothing. However, the magnitude of the unknown is, well … unknown! The physicist John A. Wheeler, who coined the term black hole, once said, "We live on an island surrounded by a sea of ignorance. As our island of knowledge grows, so does the shore of our ignorance."[10] Francis Crick, one of the two scientists who discovered DNA, used to say that a theory that fits all the facts is bound to be wrong, as some of the facts will be wrong. So don't put all your eggs in Darwin's basket.[11]

In short, what Darwin did—and what made him revolutionary—is that he approached all aspects of life as natural phenomena, which are to be explained by natural causes embodied in objectively testable theories. Thus, modern biology was born. I consider this a great part of Darwin's legacy, but it is probably not the end of the story. Only time will tell.

NOTES

1. As to the sixth edition of 1872, the title was changed to *The Origin of Species* (New York: Macmillan., 1962).
2. In *Exercitationes de generatione animalium* (1651, London) where he actually says: "omnia omnino animalia, etiam vivipara, atque hominem adeo ipsum, ex ovo progigni."
3. Although he did not come up with the concept of natural selection, Darwin's grandfather, Erasmus Darwin, did wrestle with the question of how one species could evolve into another. He also talked about how competition and sexual selection could cause changes in species.
4. Theories usually don't get accepted immediately but remain controversial for a while. William Harvey never saw his theory of a closed blood circulation being fully accepted during his lifetime. Isaac Newton's gravitation theory required some eighty years before it was universally adopted. Alfred Wegener's theory of continental drift was embraced only fifty years later, after the acceptance of the theory of plate tectonics. For almost 35 years, Mendel's findings escaped notice. And it would take decades for antiseptics—discovered by Ignaz Semmelweis (1847), Louis Pasteur (1862), and Joseph Lister (1867)—to become generally accepted in surgery. Even Einstein's 1905 work on relativity remained controversial for many years, but was adopted by leading physicists, starting with Max Planck. And the list could go on and on.
5. Don't confuse these two. Heredity is passing of traits to offspring, but this can be a matter of either genetics or upbringing. Some traits may be hereditary but not genetic (probably humor, religion, and the like). Some others may be genetic but not hereditary, because they are the result of mutations in a specific cell line of the organism (unless it is in the cell line of reproductive cells). For natural selection to have effect, the only traits that qualify are hereditary as well as genetic at the same time.
6. In this book, I won't distinguish between natural and sexual selection, as they both affect the genetic contribution organisms make to the next generation. While natural selection works more indirectly (by increasing the chances of survival), sexual selection works more directly (by increasing the chances of mating through rivalry of males and preferences of females). In selection, the number and quality of descendants depends on *both* partners.
7. Hugh Montefiore, *Probability of God* (London: SCM Press, 1985)
8. Richard Dawkins, *The Blind Watchmaker* (New York: Penguin Science, 1986).
9. Charles Darwin, *The Origin of Species*, chapter 4.
10. John A. Wheeler, *Scientific American* (1992), Vol. 267. Quoted in Clifford A. Pickover, *Wonders of Numbers* (New York: Oxford University Press, 2002), 195.

11. Facts are very sophisticated entities. Facts are always based on *interpretations* – pictured in our thoughts and expressed in statements about rock-solid events in nature. But if this is true, then there is no (scientific) factual information *without* interpretation – hence, "no interpretation" would mean "no information." The more interpretation we inject, the more information we may provide, but also the more we need to prove. Since facts are interpretations of events, there is no such thing as a neutral, clean-shaven way of seeing or observing. The command "Observe!" does not make any sense, says Karl Popper, since no one would know what to observe; theories just don't spontaneously emerge from observation. In observation, one is both a passive "spectator" and an active "creator" at the same time.

Chapter Two

Selection Filter? A Not-So-Good Legacy

The question remains, though, whether Darwin really succeeded in achieving his goal in explaining evolution with purely natural causes and laws. The answer is yes and no. He certainly came up with a scientific, natural explanation, and in that sense he did succeed and proved to be a good scientist. He was always very well aware of the fact that science should search for a lawful evolutionary mechanism similar to those mechanisms many astronomers and physicists had come up with before him in their respective fields. He explicitly followed the philosopher John Stuart Mill's rule (1872) that says "it is a law that every event depends on some law."[1] As Darwin put it in one of his letters, "astronomers do not state that God directs the course of each comet and planet."[2] What a great insight Darwin had: Comets and planets do follow natural laws—and something similar must be the case in evolution.

In another sense, though, Darwin did *not* succeed, as he ignored the following question: Why do all events depend on some law? Where do those laws come from? Why do certain biological designs "work," and why are they "successful" and "effective" in reaching their "goal"? What is it that makes them "goal-directed" to different degrees? Or put differently: What carries them through the filter of natural selection? True, these are not physical but meta-physical questions, located in a "meta" realm beyond, behind, beneath, outside, or underlying the regular physical realm—but that is not to say those questions don't exist.[3] Yet, as a scientist, Darwin decided to leave such questions untouched.

It's here that *teleology* comes in again. Teleology is about goal-directed phenomena. There is teleology in the biological world because the animate world is design-like—just as there is teleology in technology, the technical world of designers, because the world of technology is design-like as well. A heart is "for" pumping as much so as a pump is "for" pumping. However, there is also a fundamental difference between these two worlds. In the technical world, the goal of a design is also a purpose in the designer's mind. In the biological world, on the other hand, the goal of a design is not a purpose in someone's mind, not even in some animal's "mind"; caterpillars, for instance, achieve a goal without having a "purpose in mind"; they were just born that way. How come they were born that way? There are two answers to this question: In terms of *proximate* causality, genes made them that way. In terms of *ultimate* causality: They are this way because natural selection promoted such a biological design, as it was functional and well-designed in the environment they live in. With this second answer, we are back at teleology. But as I said before, Darwin rephrased teleology from an "a priori drive" to an "a posteriori result." So what is the problem?

Let me mention first that most biologists feel rather uncomfortable with the idea of teleology. The physiologist Ernst Von Brücke was right on target when he said that biologists treat teleology like a lady they cannot live without, yet they are ashamed to show themselves with her in public.[4] And it's this very teleology that leaves us some thorny meta-physical questions. Even if we acknowledge that a heart works because it is designed like a pump, and that an eye works because it is designed like a camera, we still have to face some philosophical, meta-physical questions: How come that a pump in itself "works" at all, and that a camera "works" at all? In other words, natural selection may explain that a fine working design has a better chance of being reproduced, but ultimately it cannot explain why such a design is working so well. That's where teleology keeps coming in, for design is a teleological concept in itself.

The philosophical, meta-physical answer to these questions is that a heart and a pump do work because they follow the laws and constraints laid down in what I call the *cosmic design* of this universe.[5] The universe has an overall set of restraints harnessing individual designs and making them "fit" or "successful" to a certain degree. Without a meta-physical design in the "background," biological as well as technological designs could not work at all. The most perplexing thing about a watch is not so much that someone invented such a design but that the universe at all allows for any kind of design to work the way it works, thus making certain designs more goal-directed and more successful than others—which is a matter of teleology again.

So where do purpose and design come from then? Somehow they must have been built into nature—as some kind of all-pervasive, goal-directed architecture. Let's call it the *cosmic design*. Apparently, there is "something" in successful biological designs that carries them through the filter of natural selection. In other words, natural selection on its own cannot do the "job" unless it works within a framework of cosmic design. Without this cosmic design working "in the background," there could be no selection. Natural selection can only select those specific *biological* designs that follow the rules of the *cosmic* design (by the way, designers, engineers, and architects must do the same thing).

Darwin may have thought he could reduce teleology to mere causality, but his causality mechanism of natural selection can only work on condition that there is teleology in nature. Darwin may have believed that he took teleology out of science by explaining it in terms of a "physical mechanism"—and in a way he did so—but at the same time he left it in as an untouched philosophical presupposition. Leon Kass, M.D., chairman of the President's Council on Bioethics from 2001 to 2005, couldn't have said it better: Organisms "are not teleological because they have survived; on the contrary, they have survived (in part) because they are teleological."[6] In other words, the causality of natural selection doesn't *explain* teleology, but *assumes* it. Natural selection operates *upon* what exists but offers no explanation *for* what exists.

So Darwin explained survival and evolutionary change by using teleology, but he didn't explain teleology itself. It is the cosmic design that explains teleology by determining which designs are possible—given the laws of nature and certain specific boundary conditions—and then it "selects" those designs that fit best; it is the cosmic design that contains the laws and constraints regulating which biological designs are successful in reproduction and survival—as much so as they regulate which bridges are successful designs. This is what F. R. Tennant tried to express in his statement "The survival of the fittest presupposes the arrival of the fit."[7] The fit survive because of the order and harmony between them and their environment—which is a matter of cosmic design. In that sense, Darwin did *not* change teleology from an "a priori drive" into an "a posteriori result." Teleology is not a result but a given. Natural selection does not create teleology, but its working is based on teleology.

This takes us to a related issue. Prior to any talk of evolutionary theory, William Paley (1743–1805) had argued that something as beautifully designed as the universe must have had a Designer, just like a watch does.[8] In line with Paley's natural theology, Darwin also saw successful designs in nature, but unlike Paley, he viewed nature as something designed by the filter of natural selection during a process of evolutionary change—rather than by the supernatural handwork of a Designer. However, in either case, the results must be design-like (in the sense of well-adapted, successful, and effective), because if they were not, they simply would not work in solving the problems they face. If the eye lens, for example, did not function like a physical lens, one would not see very well. That's why a mechanistic approach in science works so well; birds and planes alike must obey the laws of aerodynamics. Let's see what this entails.

George Bernard Shaw once said that Charles Darwin threw Paley's "watch" into the ocean. Well, Shaw was wrong. What Darwin did throw away was Paley's "watchmaker," but certainly not his legendary watch; if he did throw something away, it was Paley's design-Designer, but not the design concept. Make no mistake, the artifact analogy of design is as basic to Darwinism as it is to Paley's natural theology. Since the heart is designed like a pump, it is a successful design "for" circulating blood. Apparently, Darwin did not discard design or what comes with it, teleology; after Darwin, the heart still existed "for" circulation; the cause of its existence may have been different, but its teleology was not.[9]

Somehow, Darwin learned to live with the concept of *design* as a teleological concept, because without it, he wouldn't be able to let natural selection do its work of selecting the best design for a specific "goal" or "purpose" or "end." In other words, natural selection does not explain design, but by assuming design, it can explain which specific biological designs "made" it in evolution. Whether Darwinians realize it or not, Darwinism could not exist without concepts such as design and teleology.

For now, I want to stress that natural selection promotes *optimal* designs—which are usually not perfect designs. They can actually become outdated designs in Darwin's theory—designs that lost their "goal" and became rudimentary—because natural selection works under constantly changing circumstances (a feature such as the pelvis of whales, e.g., lost its function when they moved to the ocean; and on land, the swim-bladders of fish may have changed into lungs). All that matters for an optimal design is reproductive success. Success breeds success! Of course, organisms do not calculate costs and benefits, but natural selection makes them act *as if* they do make strategic decisions. And what is it that determines the "costs and benefits"? The answer is to be found in the cosmic design.

But let me repeat it once more, an "optimal" design is usually not a "perfect" design, because there are compromises to be made, when balancing the benefits against the costs. In addition, organisms live with built-in boundaries caused by their ancestral history and many other genetic and developmental constraints—which fact puts each specific design in a straitjacket so to speak. And then there are those constraints found in the cosmic design. As a consequence, designs are often far from perfect, but they are the best we can get for the time being and given the time past. Natural selection goes for the best available design under the given circumstances; as a result, designs good for life in the Sahara may not be good for life in Alaska, and designs good for living in the trees may not be good for living on the ground, and so on. Due to genetic and hereditary differences, caused by mutations, organisms differ in design, which allows natural selection to treat them "selectively."

What is the bottom line? Darwin may have thought he could reduce teleology to mere causality, but his causality mechanism of natural selection can only work on condition that there is teleology in the cosmic design of nature. In ignoring this or even denying this, that's where Darwin's legacy is heading in the wrong direction.

NOTES

1. John Stuart Mill, *A System of Logic* (Honolulu: University Press of the Pacific, 2002), chapter V, § 1.
2. In correspondence with geologist Charles Lyell in 1861.
3. Metaphysics is a branch of philosophy concerned with explaining the fundamental nature of being and the world. The term was first used as the title for several of Aristotle's works, because they were usually placed *after* the works on physics in complete editions. The prefix meta- ("beyond") indicates that these works come "after" the chapters on physics. However, Aristotle himself did not call the subject of these books "Metaphysics," but referred to it as "First Philosophy."
4. Quoted in H. A. Krebs, "Excursion into the Borderland of Biochemistry and Philosophy," *Bulletin of the Johns Hopkins Hospital*, 95 (1954): 45.
5. Some scientists would use a different term: Johannes Kepler (in his book *Harmonices Mundi*) as well as Isaac Newton spoke of the "harmony of the universe." Robert Boyle would call it the "grand economy of the universe" (*Micrographia*, 16). Later on, Albert Einstein would still speak of "the harmony of the Universe which we try to formulate as 'laws of nature'" (Letter to A. Chapple, February 23, 1954).
6. Leon R. Kass, M.D., "Teleology, Darwinism and the Place of Man: Beyond Chance and Necessity?" in *Toward a More Natural Science* (New York: Free Press, an imprint of Simon & Schuster, 1988), chapter 10.
7. F. R. Tennant, *Philosophical Theology* (Cambridge, UK: Cambridge University Press, 1969), Vol. II, 85.
8. William Paley, *Natural Theology, or Evidences of the Existence and Attributes of the Deity collected from the Appearances of Nature* (Oxford, UK: Oxford University Press, 1802/2006).
9. This point was made very clear by Michael Ruse, "The Last Word on Teleology," Essay 4 in his *Is Science Sexist?* (Boston, MA: Reidel, 1981).

Chapter Three

Heredity: The Good Legacy

Darwin was well aware of the fact that natural selection only works if there are "inborn" *differences* in design—called varieties or variations. What is it then that causes certain designs to be the way they are? What makes them different? It must be something that is inherited or, in modern terminology, genetic. Heredity is passing of traits to offspring, but this can be a matter of either genetics or upbringing. Some traits may be hereditary but not genetic (probably humor, religion, and the like). Some others may be genetic but not hereditary, because they are the result of mutations in a specific cell line of the organism (unless it is the cell line of reproductive cells). For natural selection to have effect, the only traits that do qualify are hereditary as well as genetic. When you have all your wisdom teeth removed, for instance, your descendants will still be born with wisdom teeth, for dentists pull teeth but not genes.

So we need to explain what it is that makes some mosquitoes, for instance, DDT-resistant and others not. Nowadays, we would say that's a matter of genetics. Darwin hardly knew anything about genetics, so this question was more or less uncharted territory for him. Neither did Darwin know that mutations can change genes, and therefore also their effects. We can only surmise why Darwin didn't discover genetics, which he needed so badly for his theory. My guess is that Darwin chose the concept of a blending (vs. particulate) inheritance because his main interest was in continuity[1] whereas Mendel was more interested in discontinuity when he postulated distinct genetic factors in heredity.

Before Mendel, most people thought more in terms of "blood" ties than "genetic" ties (terms such as "blood relative," "bloodline," "full-blooded," and "royal blood" are relics of this idea). So differences in temperament, talents, social status, wealth, and power were all said to reside "in the blood."

With the rise of Mendelian genetics, however, genes were substituted for blood in the explanations.[2] People differ in ABO blood types, for instance, because their two ABO genes can carry two of the same or two different alleles, which together determine the blood type they have.

Once the upcoming field of genetics began to advance, Darwin's basic theory of natural selection became more refined. I won't discuss the details here (that's for other books to do), but the current, more sophisticated version includes a network of theories and is often called neo-Darwinism or "synthetic evolutionary theory." It still assumes a process of gradual change (step by step, generation by generation, species by species), but more so on a genetic basis (gene by gene, mutation by mutation).[3]

The new theory—which is actually a cluster of theories—is based on the following concepts:

1. The source of genetic diversity is random mutation. By the way, mutations happen more frequently than you might think; each one of us has some sixty new mutations not present in either of our parents.
2. The genetic makeup of a population changes over time. Genetics is all that matters (think of those wisdom teeth again).
3. The main force behind this change is natural selection. Natural selection works through reproductive success based on a "costs and benefits analysis," as we discussed in the first chapter.
4. Additional factors are migration, isolation, and sampling. It is here that so-called genetic drift (sampling from generation to generation) and other stochastic factors may come into play. Isolation seals off the flow of genes, whereas migration redirects gene flow.[4]

Let me explain this newer version with a simple example. DDT-resistant mosquitoes make an enzymatic protein that can accommodate a single molecule of DDT and inactivate it by adding oxygen to a chlorinated side group on its molecule. Every once in a while, a malaria mosquito has a gene mutation that is capable of inactivating DDT—perhaps as a result of a duplicated and mutated gene in the "silent" DNA section (1). From now on, the population of mosquitoes has some genetic diversity in this respect. Natural selection will probably not have much impact until the population has to deal with DDT. From then on, the ones with the mutant DNA code have a selective advantage and will transfer their code much more frequently to the next generation than the carriers of the regular code (3). As a result, the genetic makeup of the population dramatically changes (2) due to the pressure of natural selection. In time, the entire population may be resistant. Plus migration might affect a similar change in nearby populations (4).

What more does genetics tell us? Some genes happen to work in one single step. There is, for instance, a gene that produces, in one step, an enzyme (*tyrosinase*) which creates the dark skin pigment melanin. This gene can also harbor so-called mutated alleles that produce a non-functional enzyme, which then may lead to albinism (although there are also other kinds of albinism). Then there are genes that require additional steps to take effect so as to cause differences between people. The gene for the ABO blood type can hold an allele for antigen A or one for B; these alleles produce an enzyme that then creates the corresponding A- and B-antigens—which is a two-step process. Then there are genes that require many more steps, or they require the cooperation of many more genes in order to produce different end-results.

As a matter of fact, the vast majority of human diseases and other genetic traits are *multi*-factorial: They are influenced by many genes interacting with one another as well as by a vast array of signals from the environment of each cell (nutrient supply, hormones, electrical signals from other cells, etc.), and all of these together will reflect the external world of the organism as a whole (upbringing, learning, experience, culture, religion). Thus, the same mutation in a specific gene may produce very different results in you and me, depending on our surrounding back-grounds as well as our genetic backgrounds (which include all other interacting genes), as each human being has a background that is unique. Even identical twins, which carry identical genes, may turn out very differently. I would say simplicity has turned into complexity!

How do such genes work then? Since the discovery of DNA, we have a better understanding of how exactly genes operate. For a while, geneticists held on to what they literally called their "central dogma" that says:[5] : DNA makes RNA makes protein—a one-directional causal flow, with one item of code, a gene, ultimately making one item of substance, an enzyme or some other kind of protein, and all these proteins together making a body. However, these views have dramatically changed.

What is the consequence of these recent changes—not to mention of the many to come, which are obviously still unknown? First of all, the key to complexity is no longer explained by the number of genes, but rather by combinations and interactions generated by fewer units of code. So organisms must be explained as entire organisms, and not as a summation of genes.

Second, DNA can never do anything on its own. It is not even capable, as many still believe, of replicating itself into copies for new generations, as it is manufactured out of small molecular bits and pieces by an elaborate cell machinery that's made up of proteins. If DNA is put in the presence of all the pieces that will be assembled into new DNA, but without the protein machinery, nothing happens. What actually does happen is that the already present DNA is copied by the surrounding cellular machinery so that new DNA strands are replicas of the old ones.[6] The process is analogous to the production of copies of a document by an office copying machine, a process that would never be described as "self-replication."

Third, DNA does in fact not make anything, not even enzymes and other proteins, because new proteins are made by a mechanism that is itself based on proteins. The role of DNA is to provide a specification as to how amino acids are to be strung together into proteins by this synthetic machinery. But this string of amino acids is not even yet a protein. To become a protein with physiological and structural functions, it must be folded into a three-dimensional configuration that is only partly based on its amino acid sequence, as it is also determined by the cellular environment and by special processing proteins. Insulin for diabetics makes a case in point. Recently, the DNA coding sequence for human insulin has been inserted into bacteria, so as to produce a protein with the amino acid sequence of human insulin. But the sequence of amino acids does not determine the shape of a protein. The first proteins harvested through this process did have the correct sequence, but were physiologically inactive. Imagine, the bacterial cell had folded the protein incorrectly! That should take DNA off its pedestal, but it does remain one of the key players in the process of heredity.

NOTES

1. See chapter 15.

2. As to whether Mendel was really the "founder" of genetics, see R.C. Olby, "Mendel No Mendelian?", in *Hist.Sci* 17, (1979): 53-72. See also G.M.N. Verschuuren, *Investigating the Life Sciences* (Oxford, UK: Pergamon Press, 1986), chapter 8.

3. More on this in chapters 5 and 15.

4. See chapter 15.

5. Francis Crick, "Central Dogma of Molecular Biology." *Nature*, 227(1970): 561–563. More in particular, he stated that genetic "information cannot be transferred back from protein to either protein or nucleic acid."

6. DNA is a double-stranded string of nucleotides (containing A, C, T, or G). You can picture the DNA helix as a twisting ladder with rungs made of one of these four base pairs: A-T, T-A, C-G, or G-C. If you split all the pairs in half, thus cutting the ladder down the center of each rung, each half-ladder contains all the genetic information needed to rebuild a complete copy of the original— for base A will only couple with base T, etc.

Chapter Four

Heredity? A Not-So-Good Legacy

The causes behind genetic diversity that we talked about in the previous chapter have become much more detailed and intricate lately. Before Mendel, diversity was a matter of "blood," after him it became a matter of genes, and nowadays it is a matter of DNA. These days, DNA seems to hold the key to life. But something has happened to the belief that DNA holds the secret of life. I call such a belief the DNA mantra. Fans of this mantra should know better; it's not that they are ignorant but they know so much that isn't so.

What isn't so then? As we found out in the previous chapter, this "miraculous" DNA is rather powerless: It cannot replicate itself, it can't even create a protein on its own. It needs all the help it can get from "outside." Think of viruses, which are essentially pure DNA or RNA; their DNA or RNA cannot do anything until they penetrate, like a Trojan horse, the interior of a "living" cell. Instead of saying that DNA is the secret of life, we could as well state the reversed: The surrounding life is the secret of DNA;[1] without its surroundings, DNA couldn't do anything!

We may have initially thought the program for living beings is located in the genes, and then in the proteins encoded by genes and their DNA. But next we found out this DNA is "controlled" by a whole network of proteins and other cellular components, and this cellular network in turn is part of an even larger network of surrounding cells, and this larger network, the organism, is surrounded by the external world. So we also need to know the "rules" of networks that control the DNA itself.

Therefore, I would say, forget the term "DNA program." DNA is more like an archive with a set of instructions, but *what* to use from this archive *when, how,* and *where* is not determined by DNA. That looks like a big blow to the clockwork concept of DNA. DNA may not be the clockwork mechanism that makes us "tick." Why not?

Well, claiming that we are nothing but DNA—because DNA is supposed to be all there is to life—brings us into a lot of trouble. Let me mention at least four philosophical problems associated with this claim: How could DNA ever create itself? How could DNA ever replicate itself? How could DNA ever discover itself? How could DNA ever validate itself? Let's tackle these philosophical questions one by one.

Problem number one: If we were nothing but DNA, DNA would have to *create* itself. Just think about that issue for a moment; it would be like the miracle of a hand drawing *itself*; a hand may draw a picture of a hand, of course, but a hand that draws a hand cannot produce the very hand that does the drawing! The very hand that does the drawing is "more" than the hand that is being drawn on paper. Claiming differently would be a case of philosophical magic, for a cause just cannot be self-caused. A cause can never cause itself, as it would have to exist before it came into existence. It certainly would be nice if gold could create itself from nothing! But we all know that's impossible—as the saying goes, "nothing comes from nothing."

Problem number two: If we were nothing but DNA, DNA would have to *replicate* itself. A copier makes copies, but it can never make copies of itself. In other words, copying done by a copy machine is certainly not self-replication. Or take the case of a projector: It can project pictures, but it cannot project itself; self-projection is beyond the capabilities of a projector. We would need something "more" to achieve this.

Problem number three: If we were nothing but DNA, DNA would eventually have ended up to *discover* itself. If we were nothing but DNA, and some of us discovered DNA, then DNA must have discovered *itself*—which is certainly sheer magic. When Watson and Crick discovered DNA, they must have been "more" than the DNA they discovered; it was certainly not their DNA that discovered DNA. I would say that is simple logic when you think about it. How could a mirror ever mirror itself, or how could a missile ever target itself, or how could a finger ever point at itself, or how could a camera ever make pictures of itself? In order to do any of these things, we need something that "transcends" any of these items, something that is capable of doing something "more" than the item itself can achieve.

Problem number four: If we were nothing but DNA, DNA would have to *validate* itself. Claims can be true or false, but molecules such as DNA cannot be true or false. In order to make any claims, we need to validate our claims as being true, otherwise they are worth nothing. If we were nothing but DNA, this very statement that we are making here wouldn't be worth more than its molecular origin, and neither would we ourselves who are making this statement. Such claims just defeat and destroy themselves. They cut off the very branch that the person who makes such claims was sitting on. I consider that suicide!

The four philosophical problems mentioned above lead us to an important conclusion: The knowing subject, you and me, must be more than the known object, DNA or whatever—in the same way as a projector must be more than the pictures it projects. Perhaps I should say it even stronger, but possibly more controversial: How could the physical world ever be studied by something purely physical, for "physical stuff" can never be true or false?

Even Charles Darwin had some awareness of this thorny problem (good for him I would say), but unfortunately he never thought it through.[2] Here is what he said in his *Autobiography*: If the theory of natural selection comes from the human mind, one might wonder whether "the mind of man, which has, as I fully believe, been developed from a mind as low as that possessed by the lowest animal, [can] be trusted when it draws such grand conclusions."[3] And again in a letter to W. Graham in 1881, "Would anyone trust in the convictions of a monkey's mind, if there are any convictions in such a mind?"[4] I myself certainly wouldn't!

Think about this for a moment: If natural selection were the origin of all there is in life, including the human mind, it would act as a boomerang that comes back to its maker in a vicious circle, knocking out the truth of the claims of whoever launched them. How could we ever trust the outcome of mere natural selection when it comes to matters of truth? In fact, natural selection must *assume* the human mind, but it can neither create it nor explain it. On its own, natural selection would be just a powerless and useless concept, for if one can't trust the rationality of human beings, one is logically prevented from having confidence in one's own rational activities—with science being one of them, including the theory of natural selection. Obviously, if Darwinism were to make such an all-inclusive claim, it would run into serious trouble in much the same way that the all-inclusive claim of determinism does. It would become its own undertaker! The trouble of such a claim is that it cuts off its very reason for reasoning and for trusting its own rationality! That is certainly a very thorny implication. Claiming that natural selection explains all there is, is certainly not a good part of Darwin's legacy.

NOTES

1. That is how Barry Commoner worded it in his "Roles of Deoxyribonucleic Acid in Inheritance," *Nature*, 202 (1964): 960–968

2. Curiously enough, Darwin applied this insight to one's belief in *God*, but not to his own belief in *evolution*. Somehow, his own theory ended up on his "blind spot." His conclusion was that we cannot trust anything we know about God, whereas I would rather argue the opposite – that we cannot trust anything we know at all if there were no God.

3. Charles Darwin, *The Autobiography of Charles Darwin* (Cambridge, UK: Icon Books, 2003), p. 149. See also Charles Darwin, *Autobiography* (1876), in *The Life and Letters of Charles Darwin*, ed. Francis Darwin, vol. 1 (London: John Murray, 1888), 307.

4. In a letter to W. Graham, 1881

Chapter Five

Reductionism: The Good Legacy

Darwin was a master in reducing a complex problem into smaller and simpler pieces. He was well aware of the fact that it is science's task to reduce complex entities to a manageable model related to an analyzable problem—which makes for a successful way of doing research. This is what is called *reductionism*—or more accurately *methodological* reductionism. This form of reductionism tries to explain things by analyzing them into smaller and smaller parts. It is a "piecemeal" approach that scientists excel in by dissecting things into pieces. "Good" scientists are those able to demarcate their area of investigation, able to limit themselves to factors that are relevant to their object of investigation, and capable of eliminating factors that might interfere with their search by keeping them under strict control. "Dissect!" is their motto.

Much of the genius of research workers lies in their selection of what is worth investigating, thus reducing their many question marks to a manageable problem. The Nobel laureate Peter Medawar used to describe research as a real art—"the art of the soluble," as he called it.[1] Hence, whatever is not soluble (yet), is not a good object of investigation (yet). Most sciences have in fact been rather successful in solving problems because they were able to create a test-tube-like shelter in a laboratory, removed from the complexity of nature, so that the various factors under investigation can be isolated and manipulated on an individual basis. The astounding successes of science have not been gained by answering every kind of question, but precisely by refusing to do so. As a matter of fact, science purchased success at the cost of limiting its ambition.

As science typically operates with a reductionistic, piecemeal approach, it shouldn't surprise us that biologists would like to apply this technique also to the problem of evolution by dissecting it into pieces—into changes on a mutation-by-mutation basis and selective processes on a step-by-step basis. In their view, selection just keeps working step-by-step (one gene at a time, one mutation at a time, one generation at a time, one species at a time) until the entire structure has been assembled. And that is what Darwin and his successors did and are still doing: They assume that the present living world is the result of a more or less gradual change, through successive living forms, from the origin to the present, with species giving rise to new species. As a consequence, evolution is a long-winding, large-scale, but step-by-step process—often too slow for us to see changes accumulating, too gradual to locate any clear thresholds, and too massive to be easily simulated in the lab.

As a consequence, even so-called "irreducible" complexity has actually been reduced by Darwinians to step-by-step evolutionary processes of mutations and natural selection. Take, for instance, the way a modern biologist would assume a scenario of a step-by-step selection process for the intricate cascading route of human blood-clotting, involving lots of proteins.[2] Most of these proteins turn out to be related to one another at the level of amino acid sequence, which most likely reflects ancient gene duplications.[3] Since new copies were not essential for the original function, they could gradually evolve to take on a new function, driven by the force of mutation and natural selection. And indeed, some of these proteins have a long history in the animal world, whereas others appear much later. That's the power of a piecemeal approach.

It is hard to deny that science has in fact been very effective in reducing the whole to the parts it is composed of. What do I mean by this statement? Let me give you some examples. Chemistry reduces molecules to a bunch of atoms; genetics reduces an organism to a collection of genes; molecular genetics reduces an organism to its DNA molecules; evolutionary biology reduces a population to a so-called gene-pool.

Let me mention to you one specific case of this "technique" as used in population genetics. Richard Dawkins, for instance, loves to use a gene-pool model in order to simulate a process of natural selection among the different variants of a gene (alleles). The gene-pool model reduces a population to a set of alleles for a specific gene and studies what happens to their allele frequencies under the "pressure" of natural selection. What we "see" happening in the model is a change of allele frequencies. The gene-pool model has been a very helpful tool.

However, we have to be careful with the use of models in science. We might actually think that the gene is the actual unit of selection in "real life," because it seems to be the unit of selection in the gene-pool model. This approach may indeed be very fruitful, but don't ever try to exchange the model for reality itself. In reality, genes are not units of selection. The biologist Stephen J. Gould pithily rebuked, "Selection simply cannot see genes."[4] Alleles, for example, that do not show up in the phenotype of a heterozygous organism cannot be subject to natural selection. In reality, the *organism* is the unit of selection, but the model may give us the misleading impression that the unit of selection is actually the *gene*. That's where the model steers us wrong: The gene is just a criterion, not a unit, of selection; it may be at center stage in the model, but certainly not in reality.

Please be aware that science works with models—which are just simplified replicas of the dissected original, made for research purposes. That is the consequence of using a "reductionistic" approach in science. Reductionism boils down to explaining things by analyzing them into smaller and smaller parts—which has been a very successful strategy in the history of science. In essence, it isolates things and uses techniques of reducing the complexity of the original to a manageable model related to a soluble problem.

But here is also where the danger comes in: Don't mistake a model for what it represents. No parent would contend with teddy bears if they were an exact representation of the original. The only model that could ever qualify as an exact replica of the original would be the original itself; but such an "all-inclusive model" is no longer a model and would also be equally complicated as the original. So, reductionism buys clarity and some certitude at the price of curtailing reality. I would say that's the good part of Darwin's legacy, as long as it stays aware of its limitations.

NOTES

1. Peter Medawar, *The Art of the Soluble* (Portsmouth, NH: Heinemann, 1967).
2. These proteins are also called "factors"; most hemophiliacs carry a mutated version of factor VIII, some of factor IX.
3. Duplication of a gene is a rather common phenomenon in genetics. The original gene keeps its initial function, whereas the duplicated version is free to accumulate mutations until it has found a new function.
4. Stephen J. Gould, "Caring Groups and Selfish Genes," *Natural History*, 1977, 86 (12): 20–24.

Chapter Six

Reductionism? A Not-So-Good Legacy

Whenever we forget to distinguish between model and reality, we may get the impression that the model has become the newly discovered reality. And that has happened to some Darwinists. They seem to have lost awareness of the fact that everything can be squeezed into the mold of a model, but doing so doesn't make the reality of the original disappear as an illusion. The late Harvard philosopher and mathematician Alfred North Whitehead once famously remarked that while math could count the atoms in an apple, it could not provide the apple itself.[1] Even a mathematical model can never replace the original. Models just give us a superficial view of the original, and only one of the many views besides.

What happens when we forget to tell model and reality apart? We end up saying that an organism is "really nothing but" a collection of genes, that a human being is "really nothing but" a string of DNA, that human values are "really nothing but" the outcome of natural selection, that altruism is "really nothing but" hidden self-interest, that thinking is "really nothing but" a series of firings along neurons, that the unborn baby is "really nothing but" a blob of cells, and that a human being is "really nothing but" a mathematical or statistical number—in other words, that the original is "really nothing but" one of its models. Sounds familiar?

Some scientists have even tried a different strategy, yet with the same outcome. They state, for example, that physics always has the last word in observation, for the observers themselves are physical. But why not then say that psychology always has the last word, because these observers make for very interesting psychological objects as well. Either statement is nonsense; observers are neither physical nor psychological or whatever, but they can indeed be studied from a physical, biological, psychological, or statistical viewpoint—which is an entirely different ball game.

C. S. Lewis coined this kind of fallacy as "nothing-buttery."[2] You will hear this myth very often when scientists leave their own territory, their own little area of expertise, and blow it up to immense proportions, thus becoming tiny philosophers without any schooling in that field. Just like some movie stars, some scientists abuse their fame to also express their worldviews—or lack thereof—thoroughly convinced that their professional or scientific expertise guarantees the quality of any other beliefs they have. Don't let plumbers advise you on your electrical wiring.

Neglecting what's outside your scope may be a wise scientific strategy, but *denying* what's outside your scope goes one step farther, actually one step too far, turning into an unwarranted ideology. Yet, that's exactly what "nothing-buttery" does; it declares "the real thing" as "nothing but …" the simplified model or map. It's like stating that highways don't exist because they are not found on our railroad maps. Science can never transcend or surpass itself by saying that science is all there is in the universe, for such would be a claim made from outside science's territory—and an illegitimate claim, besides. Isn't there more to life than science? If human beings were indeed "nothing but" DNA, they would be pretty fragile creatures; and even this very claim would be worth nothing more than the DNA that supposedly produced it.

As we discussed before, (methodological) reductionism tries to explain things by analyzing them into smaller and smaller parts. It is a "piecemeal" approach that scientists excel in by dissecting things into pieces. The reductionists' creed says: After dissecting them, things are much simpler than they appear. However, even if reductionism is a helpful scientific technique, we should still add a few footnotes. I would advise reductionists not to confuse "smaller" with "lower" (at an ontological level) and/or "simpler" (in the sense of easier to understand). I doubt whether an atom is at a "lower" level or of a "simpler" nature than a molecule; smaller parts are not always really simpler, as quantum theory has shown us.

My second concern regarding this technique is that reductionists tend to claim—without giving us any justification—that explanations should always go one way, in the direction of "smaller," "lower," and "simpler" things. But if the "smaller" can explain the "larger," why couldn't the "larger" explain the "smaller" as well? Would it really be "better" to explain the working of DNA in terms of its (smaller) molecular parts than in terms of its (larger) surrounding cellular structure? I doubt or at least question such a claim. The most we can say is that going after the smaller pieces is perhaps the easier strategy.

But there is a much bigger problem with reductionism that many scientists are not aware of. It is certainly not a problem to use this (methodological) *technique* in science, for it has been a very successful tool. But it becomes a problem when we turn it into a (philosophical or ontological) *world-*

view. If we are really—that is, in reality—nothing but DNA, we are worth nothing more than a fragile molecular structure, and so is all we claim to know. If we are really nothing but the product of natural selection, all our scientific knowledge is worth nothing more than a good fit or a best guess. Or the reverse, if we want to accept the reliability of our biological knowledge, including the theory of evolution and natural selection, we cannot conclude at the same time that all human knowledge is just a product of evolution and natural selection. First we take rationality seriously, and then we break it down with rational means. If we use the "saw" of the evolutionary theory to cut off the "branch" of the human mind, we are in fact cutting off the very branch that we were sitting on. We have actually lost any reason for reasoning and doing research. That's mental suicide!

Reductionism as a worldview denies any validity to other factors shaping our lives—factors such as motives, reasons, purposes, goals, and values. However, neglecting those factors doesn't entitle one to deny or reject them. That would be akin to rejecting everything that cannot be tested with a thermometer! Never confuse a scientific "map" of the world with the world itself; maps are only a surrogate for "the real thing," so don't replace the world with one of its maps.

That's where Darwin's legacy might easily go wrong. Darwin had at least a glimpse of this problem, but many of his followers seem to be unaware of it. When Darwinism is no longer aware of the limitations of its reductionistic approach or when it teams up with the idea of "nothing-buttery," Darwin's legacy turns into something that is "not-so-good" anymore. It becomes a dogma that demands absolute authority.

NOTES

1. Alfred North Whitehead, *Adventures of Ideas* (New York: Macmillan, 1933), 161.
2. C.S. Lewis, *The Weight of Glory and Other Addresses* (New York: Macmillan, 1980, rev. edition), 71–72.

Chapter Seven

Unified Theory: The Good Legacy

The hypothesis of evolution could be described as the result of a mental leap based on many "facts" found in nature. It does what any kind of hypothesis is supposed to do—it unifies a diversity of facts—in this particular case by assuming that all forms of life have parents, and therefore, are based on common descent with modification. In doing so, the theory of evolution makes sense of the many facts that we know from paleontology, embryology, morphology, biochemistry, genetics, and so on. The late Harvard biologist Ernst Mayr put it this way, "The theory of evolution is quite rightly called the greatest unifying theory in biology."[1] Isn't that exactly what a scientific hypothesis is meant to do—unify?

That's why the late geneticist Theodosius Dobzhansky from Columbia University could say, "Nothing in biology makes sense except in the light of evolution."[2] The diversity of organisms, with their similarities and differences and their patterns of distribution, would make for a bewildering chaos of disparate facts … until unified and given "sense" by the theory of evolution. Science is a perpetual search for unity in the midst of seeming diversity, for order in the midst of seeming chaos, for regularity in the midst of apparent irregularity.[3]

How strong is this unifying theory? you might ask. Claiming that evolution is only a hypothesis may be technically correct, but such holds for any kind of scientific theory, and is therefore not very relevant to our discussion. As a matter of fact, everything seems to indicate that evolution has indeed "evolved" from a hypothesis into a fact. Isn't it funny how we have no problem saying that our kids must have gotten certain genes from their grandparents, and yet it doesn't come easily for some of us to acknowledge that our kids also did receive some "mammalian" genes from long gone mammalian ancestors. If the idea of ancestry unifies humanity, why would it not do the same for the entire living world? Genes and their DNA in fact connect us with the origin of life through a long line of common descent. Somehow we have to accept the fact that, when compared with the rest of the animal world, we are "bone of their bone, flesh of their flesh, and blood of their blood"—albeit with much modification.

What are the facts then that wouldn't make sense if we didn't accept evolution? What are the facts that the theory of evolution unifies in one concept? Here is a short overview of several biological disciplines that have benefitted from Darwin's idea.

1. Taxonomical facts. Taxonomy is the science of classifying living things. How would we classify human beings in the living world? It may sound unpleasant, but a human being is, first of all, a biological animal; this does not mean that a human being is merely an animal, since it is an unusual, very peculiar animal, endowed with rationality and morality.[4] Nevertheless, human beings are animals, having characteristics all animals share. Just like all other animals, we breed, feed, bleed, and excrete; we are flesh of their flesh. And when narrowing down the range by zooming in further, we share even many more features. Humans also have a skeleton like some of the animals, the vertebrates, do. In addition, they start their lives in a womb, just like some of the vertebrates, the mammals, do. Then they have a relatively large brain, which they have in common with a subgroup of the mammals, the primates. And like some of these primates, the hominids, they lack a tail. When tightening the range, the resemblances become more and more numerous and striking.

Therefore, we should somehow acknowledge that we all belong to the same animal family, which would turn animals such as dogs, cats, and apes into some kind of "relatives." Does this also entail that we share common ancestors, you might ask. Not necessarily so, but if we did, this would make much more sense of the way we fit into a classification of the animal world. Human beings do have some likeness to all animals, but more likeness to animals that are vertebrates, even more so to those animals called mammals, and still more so to animals that are primates, particularly the hominids. So we seem to have many "relatives" in the animal world, but some look like close relatives and others like more distant relatives. Somehow this is a first—and I would say pretty strong—indication of common descent and ancestry.

2. Morphological facts. Let us take mammals as an example of close morphological similarities. All mammals share more or less the same morphological structure, in spite of very obvious differences in appearance. Just the fact that all mammals have consistently seven cervical vertebrae illustrates this point—no matter whether it is in the sturdy neck of a rhino or the long neck of a giraffe. They also share the same basic pattern in their extremities, but then adjusted to specific needs of flying, digging, swimming,

running, etc. Some common basic structures may have become "overdeveloped" when it was beneficial; others seem rather "underdeveloped," once they had lost their functionality in certain groups of mammals. Structures that have lost their functionality are called *rudi mentary*; such "left-over" parts only make "sense," if we assume that all mammals derive their same build from common ancestors.

Instances of rudimentary structures are all over the mammal world. Whales have an under-developed pelvis, but they still do have a tiny pelvis dating back to when they were land mammals. We ourselves share ear muscles and an appendix with other mammals, but these have lost their original function. As a matter of fact, all mammals (including humans) still have floating ribs and vertebral processes that were most likely real ribs in their long-gone ancestors. And in the evolutionary tree of horses, we see certain parts of their hoofs gradually disappear while other parts have become very prominent. And all of this is just a small, rather arbitrary selection of morphological evidence for common descent. Let me repeat it again: These "odd" phenomena only make sense, if we may assume that all mammals have common ancestors and inherited the same body plan. Darwin worded it as follows: "The framework of bones being the same in the hand of a man, wing of a bat, fin of the porpoise, and leg of the horse,—the same number of vertebræ forming the neck of the giraffe and of the elephant,—and innumerable other such facts, at once explain themselves on the theory of descent with slow and slight successive modifications."[5]

Even in animal *behavior*, we may find traces from evolution's past. Amphibians always must return to the water to lay their shell-less eggs, even those who have become land animals (such as terrestrial salamanders). Reptiles, on the other hand, must lay their eggs on land, even if they spend most of their time in water (such as sea turtles). Their evolutionary past seems to have taken its toll on the way they behave.

And then we have those weird nipples on the chest of human males. Where might they come from? Regardless of their sex, humans in the womb look alike until sexual hormones start differentiating males from females. Due to common ancestry with other mammals, all humans have a line of nipples appear along the two vertical "milk lines," which start in the armpit on each side, run down through the typical nipples, and end at the groin. During sexual differentiation of humans, most of these nipples disappear before birth, except for one pair; but in some cases (5%) a few of the other nipples remain visible and are often mistaken for moles. The regular, remaining single pair develops rich mammary glands due to the effect of female hormones, but this may also happen to males if there is a hormonal imbalance. Therefore, males have nipples because females do; and females have nipples because all mammals do. That's what common descent seems to have done for us.

3. Embryological facts. During the embryological development of organisms, we often find intermediary stages that were once, in their ancestors, final stages. In general, if a structure originated before another structure in evolutionary terms, then it also appears earlier than the other in the embryo. Human embryos, for instance, do develop "gill" pouches just like fish do, but they never become real gills; and the human cerebrum develops last. Accordingly, if a structure disappeared during evolution, then a corresponding structure may appear first but then disappear or become modified. Whales, for example, have no more legs, but during embryonic development, leg extremities do occur, but then recede; they also have hair at one stage, but lose most of it later.

In other words, when you compare the embryological development of various mammals, the beginning stages usually show a greater deal of similarity than the end stages, which may differ dramatically. Thanks to those shared beginning stages, we may often be able to trace their ancestry. For example, the fact that pharyngeal "gill" pouches appear briefly during the early embryology of mammals, including humans, seems to indicate that the distant ancestors of land vertebrates were aquatic; our ancestors must have come from the water (even though apes can't swim anymore). And then we see that human embryos still develop a layer of downy hair, but they shed this layer at about 35 weeks of gestation. We could also say that evolution has somehow left its "imprint" behind. In Darwin's own words, "we can clearly see why the embryos of mammals, birds, reptiles, and fishes should be so closely alike, and should be so unlike the adult forms."[6]

By the way, what I just said does not entail by any means that human embryos as they develop would be retracing the whole evolutionary history of humanity, step by step. The human embryo never has gills in any sense of the word. The fanciful notion of gills is based upon the presence of six alternating ridges and grooves in the neck region of the human embryo. While similar arches do give rise to gills in certain aquatic vertebrates such as fish, their developmental fate in mammals has nothing to do anymore with gills or even breathing. In humans and other mammals, these arches and pouches develop into part of the face and its jaws (1st arch; a faulty closing of the 1st arch would cause a cleft palate), muscles of facial expression (2nd arch), some endocrine glands (3rd and 4th), and the larynx (6th), whereas the 5th arch even vanishes. So, a human embryo does not become human at some point during its development, but rather is uniquely human at every stage of its development—which is partially due to the fact that its DNA is human from the very start.

4. Mimicry facts. Another instructive case comes from the study of mimicry as it was started by the English Darwinist H. W. Bates (1825–1892). He noticed that signals from an animal species that is effectively protected by being poisonous or distasteful may be mimicked by an animal species that is

not naturally protected from predation by some unpleasant character of its own—which is something that some "harmless" flies (without a sting) do by merely looking like really stinging wasps. This means that the unprotected species takes advantage of its *visual* resemblance to a protected species. Interestingly enough, Bates discovered that when the protected species varied geographically, its mimicking satellites had undergone exactly the same changes as their unpalatable models. Evolutionary theory could easily explain this.

5. Paleontological facts. Paleontologists have unearthed an immense collection of fossils, most of which come from a flora and fauna very different from ours. In general, the deeper the layers from which they originate, the more their fossils differ from their current counterparts. Mammals and fruit plants, for instance, only occur in higher, more recent layers (which can be independently dated by the technique of radioactive dating of rocks). In general, fossil formation has a better chance of happening in water (away from oxygen in the air), as bones are buried by sediment, sand, and other materials that settle out of the water. In time, the layers of mud and sand become covered by more deposits, thus pressing down the layers holding the fossils and then hardening them into rock. Later on, changes in the earth surface may expose parts of these layers again when stream erosion wears away layers of sediment.

As to many types of animals, so-called evolutionary trees have been constructed, showing the hypothetical connections between related fossils. In addition, we can still find existing intermediate forms between fishes and reptiles (such as amphibians, who grow up in water like fish, move to land like reptiles, and come back to water to lay their eggs), or between reptiles and mammals (think of so-called living fossils such as the duck-billed, beaver-tailed, otter-footed mammal called *Platypus*, known to lay eggs developed in a "mammalian" uterus). But we also did unearth fossils of transitory stages, such as fish-like reptiles, reptile-like birds, and reptile-like mammals. The particular shape of evolutionary trees may be up for discussion, but the tree itself is more or less a "scientific fact." However, since fossilization is a relatively rare phenomenon, many connections may be missing, producing those so-called "missing links." That's where inevitably some guess work comes in, which causes a more detailed tree to be less scientific than we perhaps had hoped for. Fortunately, nowadays we have newer and better tools at our disposal to assess evolutionary connections in more reliable and more scientific ways, as we will discuss soon.

Additional evidence comes from the fact that the current location of continents is allegedly the result of a drastic transformation. Originally, all continents formed one large continent (*Pangaea*), but parts of it drifted away from each other to reach their current position.[7] Australia, for example, had been isolated already from the other continents before mammals appeared on stage; consequently, the marsupials could flourish in the Australian fauna by lack of mammalian competition, making them look like their ecologically equivalent mammals, who are widespread in the rest of the world but are not native to Australia. Phenomena like these only make "sense" if we assume that isolated continents and regions had their own evolutionary processes, independent from other areas. It would be very hard to explain these "odd" facts without the assumption of some kind of evolutionary process.

6. Chromosomal facts. Chromosomes form at the time that a cell divides into two new cells, which makes them visible through the light microscope. They are packed with DNA (more on DNA under #7), and may contain hundreds of genes. Each species has a specific number of paired chromosomes—one half coming from the female parent and the other half from the male parent. Chimpanzees, for instance, have 24 pairs of chromosomes, whereas humans have only 23 pairs.[8] This change from 24 to 23 pairs of chromosomes can be explained if we assume that two ancestral chromosomes have fused together to form the human chromosome #2. This is even more likely so since gorillas and orangutans also have 24 pairs, just like chimpanzees.

Is there any more specific proof that chromosome #2 is a head-to-head fusion of two shorter chromosomes? Yes, there is. The tips of all primate chromosomes have a special sequence of DNA code which is rare elsewhere. Well, this very sequence is found right in the middle of our fused chromosome #2. The fusion has left its DNA imprint behind, so to speak, making it hard to explain this phenomenon without postulating common ancestry and descent.[9]

7. DNA facts. DNA is packaged into chromosomes and contains the genetic information that steers the synthesis of proteins. Protein synthesis takes place in cells and typically goes through the following steps: DNA in the cell nucleus acts as a template for synthesis of mRNA; mRNA leaves the nucleus and goes to the cytoplasm, where it complexes with ribosomes; tRNA carries one specific amino acid to mRNA; tRNA couples briefly with mRNA on a ribosome; the ribosome moves along on mRNA, adding amino acids to the growing polypeptide chain resulting in a protein; tRNA moves off to pick up more amino acids.

All the steps outlined above present actually a very simplified version of the full story. Some DNA sections, so-called *introns*, get initially transcribed into mRNA but are then removed from the end-product by so-called splicing. Due to alternative splicing, a single gene may code for different protein sequences. Introns may also contain "old code" that has become inactive. New discoveries are being made on an almost daily basis, constantly revolutionizing our current understanding of genetics.

Since just about all organisms contain DNA, comparing the DNA structure of different species may be an indication as to how closely they are related.[10] DNA is like a logbook reaching almost as far back as the origin of life. Because of common descent, our human DNA actually connects us with a very long line of ancestors with whom we share parts of our DNA.

The one-time idea was that DNA holds genes coding for either structural proteins or enzymatic proteins. This idea has been refined by the discovery that protein-coding regions of genes can be interrupted by DNA segments that play more of a regulatory role by producing activator and repressor proteins that either activate or repress the activity of "regular" genes. Some of these regulatory genes are actually very short and do not produce proteins at all but short strands of mRNA capable of blocking the mRNA of a "regular" gene from creating its protein; they are called micro-RNA-genes.

In addition, genes may be separated by long stretches of DNA that don't seem to be doing much—that's why they are often called "junk DNA," but this designation may turn out to be too premature, betraying our lack of knowledge; "non-coding," "neutral," or "silent" DNA would be a safer term. Some of this "non-coding" DNA is repetitive DNA, often replicated from regular, coding DNA, and perhaps a rich source from which potentially useful new genes can emerge in evolution (if so, this DNA may not really be "unemployed").

As it turns out, DNA segments of regular genes ("coding" parts) are more stable during evolution than DNA segments in between genes ("silent" parts)—probably so because they are more sensitive to harmful mutations. We share, for instance, 100% of our genes with chimpanzees, but only 98% of the DNA in between those genes; with fruit flies, however, we share only 60% DNA inside genes and as little as 0% DNA located in between genes.

Comparing DNA from different species gives us a rather strong indication of common descent between organisms. But there is even better evidence based on so-called *pseudo-genes*. These are genes that resemble a regular DNA packet of a functional gene, but they have been affected by one or more glitches that change their script into "nonsense." They were once functional copies of genes but have since lost their protein-coding ability (and, presumably, their biological function). In a way, pseudo-genes are like rudimentary organs (see #2). When comparing humans and chimpanzees, we do find genes that are functional in one species but no more in the other (so we call them pseudo-genes). A striking example is the gene for a jaw muscle protein (MYH16), which has become a pseudo-gene in humans, but is still very functional to develop strong jaw muscles in other primates. Another example would be the DNA sequence for an enzyme that produces ascorbic acid (vitamin C) in most animals. Many primates, including humans, have a defect in this DNA code, so they must acquire vitamin C through food, but they did hold on to its repetitive sequences in the "silent" section of their DNA.

I think I have shown you enough how nicely and beautifully the theory of evolution and natural selection unifies a whole gamut of biological phenomena and facts into one theory. All of the above facts are hard to explain unless we assume some kind of evolution; they just don't make sense if we do not assume evolution. The theory of evolution really qualifies as a showcase of "grand unification" in science. It is a nice example of a "grand unified theory" in biology. Thank you, Darwin, for this great legacy!

NOTES

1. Ernst Mayr, *Populations, Species, and Evolution* (Cambridge, MA: Harvard University Press, 1970), 1.
2. Theodosius Dobzhansky, "Biology, Molecular and Organismic," *American Zoologist*, Vol. 4 (1964): 443–452.
3. Einstein's work is a good example of this: "The simpler our picture of the external world and the more facts it embraces, the stronger it reflects in our minds the harmony of the universe" (Albert Einstein and Leopold Infeld, *The Evolution of Physics*, CUP Archive, 1971, p. 213).
4. See chapter 17.
5. Darwin, On the Origin of Species (London, John Murray, 1859), chapter 14.
6. Darwin, *On the Origin of Species* (London: John Murray,1859, 1st edition) 1st issue, 497.
7. This is called *continental drift* based on the plate tectonics theory.
8. The number of human chromosomes is actually 22+XX for females or 22+XY for males.
9. By the way, fused chromosomes are not uncommon; the domestic horse (32 pairs), for example, has a fused chromosome pair compared with the *Prezwalski* horse (33 pairs).
10. The technique to do so uses DNA from two different species and makes its strands uncouple. Next, single-stranded DNA from the two organisms is mixed under conditions that make them re-couple. The amount of coupling gives us a measure of the degree of genetic similarity between the two species.

Chapter Eight

Unified Theory? A Not-So-Good Legacy

The idea of "grand unification" or a "grand unified theory" has always been very appealing to scientists. Physicists, for example, are still in search of a physical "umbrella" theory that would unify all physical forces in the universe. Let me briefly explain this idea. All scientific laws operate within so-called boundary conditions or restraints. There are half a dozen physical constants that are given, not derivable from any known theory (the speed of light, the force of gravity, etc.). The law of gravitation, for example, is constrained within the boundaries set by general relativity for a specific average density of matter—otherwise a cosmological catastrophe would be generated. All other physical factors are causally determined by these more fundamental constants. It is not impossible, though, that some of these fundamental constants may in time be derived themselves. At one point, for instance, the boiling point of water was taken as a physical constant, but is now considered the result of quantum mechanical laws. In time, the speed of light constant may be tied to Planck's constant defining the smallest possible unit of space and time.

Whatever the outcome is, a search for a physical "umbrella" theory that would unify all physical forces in the universe seems to be a valid undertaking. In a similar way, searching for a biological "umbrella" theory that would unify numerous biological phenomena through the theory of evolution and natural selection is a very laudable enterprise as well. That's where Darwin deserves big accolades!

However, the next step for many scientists, including some biologists, perhaps even Darwin himself, is to inflate their "grand unified theory of *biology*" into a "grand theory of *everything*." It is their claim that such a super-theory, if ever found, would not only explain all biological facts, but also everything else in our universe—a Grand Theory of Everything in this Universe, so to speak. Striving for such an all-compassing theory has been a timeless temptation to scientists. The atomic theory once had this disreputable aura of a Grand Theory of Everything—not only claiming that atoms are everywhere, but also that they are all there is (materialism). Later on, it became an important philosophical doctrine to declare that all living objects (including human beings) are only and merely machine-like automata (mechanicism). And the list could go on.

Something similar happened to Darwin's theory of natural selection—also called the theory of "survival of the fittest"—when another player appeared on the scene to steal Darwin's legacy: Herbert Spencer was his name (1820–1903). This self-proclaimed philosopher saw a lucrative ideological goldmine in Darwin's theory of natural selection and quickly grasped the worldview potentials of that theory by changing it into a Grand Theory of Everything. It was Spencer who applied "natural selection" to a wide assortment of phenomena, ranging from psychology, to sociology, to ethics—and he rephrased the theory in terms of "survival of the fittest."[1] Based on his new slogan "Survival of the Fittest," Spencer claimed that everything in life is based on natural selection, including our social relationships, our motives, our desires, our beliefs, and our morals. By doing so, Spencer exported the theory from the biological realm to the realm of sociology, psychology, and ethics. He wrote four books to promote his new "religion": *Principles of Biology* (1864–1867), *Principles of Psychology* (1870–1872), *Principles of Sociology* (1876–1896), and *Principles of Ethics* (1892–1893).

I would say that's when Darwinism became a secular religion with an eschatology based on evolutionary progress for the happy few—for the fittest among us, to be precise.[2] Apparently, the hypothesis of "survival of the fittest" had become the new "Grand Theory of Everything" for Spencer and his supporters.

I think it's needless to say that, in all such cases, the boundaries of the underlying theory are being grossly overstepped—thus elevating a dedicated theory to the level of a worldview, an ideology, or even a dogmatic doctrine, claiming *universal* validity for *local* successes. That's where science goes overboard. Whenever this happens, I would like to use the ending "-ism," so we would speak of ideologies such as atom*ism*, physic*ism*, evolution*ism*, scient*ism*, and so on. All "isms" tend to go overboard; they love to simplify the vast complexity of reality; they look at the world from one specific perspective and claim this makes you see everything—that is, literally, all there is. They replace reality with one of its specific models or maps.

All "isms" offer us one side of the story, but maintain there is no other side, so they all suffer from some form of megalomania and are all monopolistic ideologies that demand absolute authority. But I always wonder what gives their proponents the certainty that only their own particular ideology is certain and that there is nothing else in life. The problem is that we cannot positively assert a "universal negative"; it is much easier to establish that something has a certain cause than to prove that there isn't any cause at all. It is impossible to prove that there is nothing else beyond what is being claimed. I must qualify, though, that when taken in this sense, Darwinism per se is not an "ism," unless it inflates itself to a real "ism" that claims to explain *everything* in life. Let's face it, the doctrine of evolution does not answer all questions, and it certainly does not answer above all the great philosophical question: From where does everything come?

Besides, there is some strong mathematical proof to demonstrate that a deterministic Grand Theory of Everything is necessarily flawed. The logician and mathematician Kurt Gödel proved in his famous so-called *incompleteness theorem* that no complex mathematical theory can be both consistent and complete at the same time; if it is complete, it cannot be consistent; and if it's consistent, it cannot be complete, for its consistency cannot be proven within the system.[3] The late physicist Stanley Jaki once pointed out that any "grand unified theory" will certainly be a consistent mathematical theory, and therefore must be incomplete—which would doom all searches for a deterministic Grand Theory of Everything.[4] The proof of either its consistency or its completeness must then lie outside the theory itself, which is necessarily meta-physical territory where scientists lack competency and feel alien.

No wonder, opposition to Darwin's theory has been as old as the theory itself—not so much because of its biological implications but due to its much wider ramifications and claims. Many people foresaw enormous consequences if the theory would be exported beyond its own territory. The most famous and yet despised opponent shortly after Darwin had published his theory was probably the Anglican Bishop Samuel Wilberforce. He took on the debate with Thomas Huxley—who replaced Darwin in the debate and is often called Darwin's "bulldog"—during a meeting of the *British Association for the Advancement of Science* in Oxford on Saturday June 30, 1860.

In response to what Huxley took as a jibe from Wilberforce as to whether it was on Huxley's grandfather's or grandmother's side that he was descended from an ape, Huxley made a reply which he later recalled as being that if asked, "would I rather have a miserable ape for a grandfather or a man highly endowed by nature and possessed of great means and influence and yet who employs these faculties and that influence for the mere purpose of introducing ridicule into a grave scientific discussion, I unhesitatingly affirm my preference for the ape." No verbatim record was taken; eyewitness accounts do exist, but vary somewhat.

Bishop Wilberforce was portrayed by many in the press as an anti-evolutionist, but what those reporters usually leave out is the fact that he should be depicted first of all as a leading anti-racist who walked in the footsteps of his father, an anti-slavery activist.[5] One of the Bishop's worries about Darwinism was that it could be *used* to justify the enslavement of blacks.[6] Bishop Wilberforce could allow that evolution may be true, or that there are significant differences between the races, and still maintain that we are not just animals, and that all humans, whatever the color of their skin, are God's children.

Apparently, Wilberforce was very well aware of the limitations of science, so he had the opinion that, if taken the wrong way, Darwinism might have very widespread, but dangerous social and moral implications. So, it was actually *social* Darwinism—and its idea of "misfits" in a world of "supremacists"—that was at stake in this debate, and not so much its *biological* version.

As a matter of fact, it was this wide-ranging dogmatic Spencerian worldview that would dominate also at the famous "monkey trial" in Tennessee (1925). It has been a widespread misunderstanding that this trial was all about Darwin's scientific theories, but that's not the entire story. One of the counsels during the trial was the Presbyterian Congressman William Jennings Bryan (1860–1925). Bryan believed moral progress could achieve equality at home and, in the international field, peace between all the world's nations. In opposing the theory of evolution, he did so not only because he believed that what he considered a materialistic account of the descent of man through evolution undermined the Bible, but most of all because he considered Darwinism a form of *social* Darwinism, promoting hatreds and conflicts in the world, and leading to a new World War (how providential!). In one of his lectures, Bryan warned his audience that the theory of evolution could undermine the foundations of morality. However, he conceded, "While I do not accept the Darwinian theory, I shall not quarrel with you about it."[7]

So again, the Scopes trial in 1925 was not so much a trial about biological Darwinism as it was a tribunal on social Darwinism—that's to say, the Spencerian idea that human society is an arena of struggle in which "supremacists" should survive, at the cost of "misfits." Even the well-known Harvard biologist Stephen Jay Gould had to admit that Bryan's anti-evolution views were a result of his populist idealism and suggested that Bryan's fight was in fact aimed at *eugenics*, the social version of Darwinism.[8] Interestingly enough, the most vehement opponent of Bryan at the time was the journalist H. L Mencken, who was known for idealizing German culture, for adoring Nietzsche and his "superman" (*Űbermensch*) idea, and for making racial and anti-Semitic remarks. Again, the trial wasn't so much a "monkey" trial as a "superman" trial.

I think we must come to the conclusion that any attempt to make a "Grand Theory of Everything" out of the theory of natural selection is one of the not-so-good parts of Darwin's legacy. Darwin may not directly be responsible for such attempts, but he did sow its seeds, and those efforts have definitely tainted the further developments of Darwinism and its reputation.

NOTES

1. Darwin adopted this term as well in his 5th edition of *On the Origin of Species*.
2. See Michael Ruse, "Is Evolution a Secular Religion?" *Science*, 7, March (2003).
3. Francesco Berto, *There's Something about Gödel: The Complete Guide to the Incompleteness Theorem*. (Hoboken, NJ: John Wiley and Sons, 2010).
4. Stanley Jaki, *The Relevance of Physics* (Chicago: University of Chicago Press, 1966), 127–130.
5. Wilberforce College in Ohio, the second oldest college for black higher education in the US, was named after the Bishop's father.
6. I have to stress, though, that Darwin himself had anti-slavery convictions as well.
7. In his famous Chautauqua lecture, "The Prince of Peace."
8. Stephen J. Gould, *Bully for Brontosaurus—Reflections in Natural History* (New York: W. W. Norton, 1991).

Chapter Nine

Randomness: The Good Legacy

One of the keywords in Darwinism, and even more so in neo-Darwinism, is the word *randomness* or chance. It is a keyword that plays a very ambiguous role in the development of Darwin's theory. Darwin himself opens chapter 5 of his book *On the Origin of Species* with the following apology: "I have hitherto sometimes spoken as if the variations [...] had been due to chance. This, of course, is a wholly incorrect expression, but it serves to acknowledge plainly our ignorance of the cause of each particular variation."[1] In other words, his speaking of chance as the cause of variations is in fact simply acknowledging that they appear to have no assignable cause. As to Darwin, using the term "chance" merely signals ignorance of the true cause. Whatever the cause of the generation of a variation may be, the role of selection is to accumulate those already present variations that "happen to be" beneficial. But what he did state and stress is that they are not the result of some "purpose."[2]

It is especially in the further course of Darwinism that this keyword becomes more prominent: Mutations are supposed to be random, environmental changes are considered to be random, and natural selection allegedly acts randomly. We can easily replace the phrase "at random" with the expression "by chance," because they are basically interchangeable. However, the term chance can carry many different meanings. Each time a biologist says that something happened "by chance," we should ask: How do you mean? Only then can we start a reasonable discussion.

To begin with, mutations in genetic material are considered to happen "by chance." What scientists mean when saying this is, first of all, that mutations are unpredictable (as far as we know) as to when and where they strike. We may know more nowadays about what causes them (factors such as radiation, chemicals, free radicals, etc.), but we still cannot predict at what location in the DNA those factors will hit and what changes they might generate there. They may not be principally unpredictable but practically unpredictable. Second, they mean that mutations are unrelated, since there is no connection with anything else going on (other than radiation or chemical pollutants). Third, they mean that mutations are somehow arbitrary, because mutations do not select their target but hit indiscriminatorily. Fourth, they mean that mutations are just opportunistic, because they occur regardless of any future needs.

Then there are environmental changes; these are also considered to happen "by chance." But this time, those changes may very well be predictable (based on climatology, etc), but they do happen "by chance" in being arbitrary—that is, without any preference for a certain location. They are also undirected because they don't strive for a preset outcome.

And then we have chance as it plays a role in genetics. When male organisms produce sperm cells or when female organisms produce ova, it is a matter of chance as to which chromosome parts get exchanged (recombination through crossovers) and which single allele of each pair of genes ends up in a reproductive cell. In these cases, chance is mostly a matter of "no preference" and "no foresight." Yet, there is some predictability involved (according to Mendel's statistical laws, etc.) as well as some correlation (as genes may be linked to each other, etc.). However, "chance" is as native to Mendelism as it is to Darwinism.

What about natural selection then? Natural selection only happens "by chance" in one particular sense: It makes a population adapt to a volatile environment, regardless of the changes the future will bring. This may turn out to be a very opportunistic, short-sighted adaptation (*micro*-evolution), but in and of itself it is highly selective, and definitely not random in the sense of arbitrary. Natural selection has a definite "preference" for "the fittest." But it does seem to be true that micro-evolution does not have a direction other than instant adaptation. Let me give you an example: If air pollution makes peppered moths with a dark color resting on the bark of trees blackened by soot effectively camouflaged, a reversal of the pollution would favor the lighter variants again. Obviously, selection can go either way; if you would film the first part, playing the film backwards wouldn't strike you as strange. Sometimes, organs first develop and then degenerate; so you could play the film also backwards.

Evolution on a larger scale (macro-evolution), however, is a more controversial issue in neo-Darwinism. If *macro*-evolution is only lingering micro-evolution, it would also be an undirected, short-sighted phenomenon. Nevertheless, some biologists like to see a long-term trend in evolution toward more complexity, more efficiency, more independence, and more intelligence (in spite of seeming oscillations). Were we to play such an evolutionary movie backwards, we would say it is going the "wrong" way, something like a stream running up-hill—which means that somehow we do perceive some kind of direction in evolution. In evolution, the time arrow seems to move from whatever is less complex to whatever is more complex. Some would say that natural selection follows a certain "path" just like a river does.

No wonder that the concept of "randomness" has always been a very divisive issue in Darwinism—not only because of its many diverse meanings, but also of the connotations it may carry. The famous or infamous question that anti-evolutionists like to pose runs as follows: How could something as complex as a cell ever emerge from simple molecules by pure *chance*? And then they paraphrase this question as follows: How could a chimp pounding along on a typewriter ever produce Shakespeare's Hamlet by pure *chance*? The answer is, of course, that this is virtually impossible!

However, we should ask these opponents whether this analogy is correct and fair. First, I would point out that the outcome doesn't have to be Hamlet—any kind of literary masterpiece would do. Second, the analogy wrongly assumes that all letters of the alphabet have an equal chance of occurring—which they don't. And third, the analogy suggests that all possible combinations are equally likely to occur. The second and third assumptions are certainly not the case in language, but neither are they in nature. When carbon and hydrogen come together, they may form CH_4, but not a mixture of all imaginable combinations such as CH, CH_2, CH_3, CH_5, and so on. Somehow, the cosmic design of the universe is such that certain combinations are much more likely and stable—and that's why certain proteins may end up acting as catalysts, for example. When a dice constantly throws a six, we would say it must be loaded. Well, creation seems to be "loaded" too—loaded with cosmic design. Therefore, I would warn you for these kinds of deceiving anti-evolutionist analogies.

To sum up, randomness in biology stands for one or several of the following statements: the outcome is unpredictable, and/or unrelated to other causes or effects, and/or arbitrary without any preferences, and/or opportunistic without any foresight, and/or undirected without a certain preset outcome. Many processes in evolution possess one or more of these characteristics, but no matter what their outcome is, when it comes to natural selection itself, we are dealing with a process that is certainly not arbitrary but preferential—although it may be a short-term kind of preference focused on instantaneous needs. As we found out, natural selection promotes good designs at the cost of bad designs by favoring biological designs that are functional in the way they solve a problem posed by the environment. So the more an organism is adapted to its environment—making for a better design fit to solve a problem posed by the current environment—the more likely this organism is to contribute to the genetic constitution of the next generations. There isn't much randomness in this part of the story.

But indeed, there is short-sightedness. That's how Darwinism can deal with "flaws" in evolution, because it considers them to be evolutionary "compromises" as well as seemingly opportunistic, short-term solutions that don't seem to be the perfect adaptations we might expect. The accepted paradigm *before* Charles Darwin hypothesized a perfect adaptation of organisms to their environment under a Creator's design. Non-adaptations were anomalies in this framework. It was Darwin who pointed out that structure and function may not always be in agreement. That's why such anomalies went unrecognized until Darwin put forward his new theory of natural selection. In a world of instant designing, non-adaptations are anomalies, but in a world of natural selection, they are not: A trait that was formerly beneficial—such as the wings of ostriches and penguins that do not fly, or the webbed feet of upland ducks which rarely go near the water—may still be inherited while no longer beneficial. Environments happen to be volatile.

Hence, evolution is full of non-adaptations. The appendix is the remnant of an organ once very functional when our ancestors were still plant-eaters. And our jaws, which are not so protruding anymore as they used to be in our ancestry, have become overly crowded, often causing the third molar to be in trouble (our "wisdom teeth"). And it's because the human birth canal passes through the pelvis (which is also a structure inherited from our ancestors), that the baby cannot be born naturally if the baby's head is significantly larger than the pelvic opening (which calls for a "caesarean section"). Or take the case of the human male, whose testes develop initially within the abdomen (being another ancestral feature dating back to "cold-blooded" animals), but then later on, during gestation (although sometimes delayed), migrate through the abdominal wall into the scrotum (as is typical and beneficial for "warm-blooded" animals); no wonder this process causes two weak points in the abdominal wall where hernias can later easily form.

As a consequence, Darwinism takes biological designs always as optimal designs (the best of what's available), but not necessarily perfect designs, let alone intelligent designs. An "optimal" design is usually not a "perfect" design in evolution, because there are compromises to be made, when balancing the benefits against the costs. Plus organisms live in changing environments. In addition, they live with built-in boundaries caused by their ancestral history and many other genetic and developmental constraints—which fact puts each specific design in a straitjacket so to speak. And then there are those constraints found in the cosmic design. As a consequence, designs are often far from perfect, but they are the best we can get for the time being and given the time past.

Take the example of the retina in the human eye. It is often mentioned as an ideal example of perfect design, but it is actually structured "inside out," with the nerves and blood vessels lying on the surface of the retina (instead of behind it, as is the case in many invertebrate species); as a consequence, this arrangement creates a blind spot and calls for a number of complicated adjustments, because light has to pass through nerves and blood vessels first in order to reach the light sensors. I wonder whether this can be considered good design from the viewpoint of a designer. And what to make of the fact that so many people have to live with back pains? Although humans stand on two legs, their skeleton was originally designed for four. Evolution accomplished this transition by breaking up the bow-curved arch of the back of four-legged animals into the s-curve we as bipeds now possess. The result is a set of creative adaptations, but with compromises. Back problems are the "scars" of this evolutionary past: strained muscles, slipped discs, and pinched nerves. Another outcome is that the long and flat pelvis of quadrupeds has become more basin-shaped with a short, broad top bone, good for balance but bad for birthing because it narrows the birth canal.

Again, selection goes for optimal, yet sometimes shortsighted, designs based on compromises and harnessed by existing constraints. Randomness is part of that explanation. Randomness comes with Darwinism as it takes evolution as some sort of "mechanistic" process that has no goal, no purpose, no direction, no preference, and no foresight—yet it creates some pretty good biological designs. To think differently would be a rationalization beyond the reach of science anyway.

NOTES

1. Darwin, *On the Origin of Species*. (1859) London: John Murray. 1st edition, 1st issue, 149.
2. See chapter 11 on this thorny issue.

Chapter Ten

Randomness? A Not-So-Good Legacy

As I said in the previous chapter, one of the main reasons why Darwinism has been under vehement attack is its emphasis on randomness. Often those attacks are based on a confusion of different meanings behind the terms "randomness" or "chance," as we discussed in the previous chapter. But there is a much more important reason why Darwinism has become the target of anti-evolutionists and that is the fact that some Darwinists would like to take the word randomness to an entirely different level, which is the level of a worldview. They take chance or randomness to be another name for some kind of capricious, blind agent called fate. I would call this "chance with a capital C"—the goddess of Fate or Doom or Blind Fate.

Let's put it clearly: Science has nothing to say about chance with a capital C. Fate is far beyond its reach; fate is in essence a worldview notion. At the very moment we get to this particular meaning of chance—in the sense of "being meaningless and without a purpose"—we have left the territory of science. No matter how selection operates, chance with a capital C is absolutely beyond its reach and expertise. The outcome of the next generation can never be declared a scenario that is orchestrated by the goddess of Fate; science can never proclaim that one is genetically "doomed" or "favored" by blind destiny, for that would make science a semi-religion. Don't preach science, teach it!

What I want to stress in this discussion is the fact that the concepts of chance and randomness in evolutionary theory never even come close to metaphysical or religious concepts such as fate, destiny, doom, gloom, or meaninglessness. Those who take the word chance out of its scientific context and interpret it as meaning "senseless" and "meaningless" change "Chance" into a capricious, blind agent, actually a deity, or a "blind watchmaker" at best, that turns the whole of life into a mere play of whimsical and fortuitous events. After doing so, they claim that evolution is a senseless and meaningless process—which is a dogmatic worldview, nicely disguised in the garment of science. Science has thus been turned into a pseudo-religion: omni-potent and omni-present.

Then there are other scientists who consider randomness to be the opposite of order. In essence, they also turn "Chance" into a capricious, blind deity, but this time, it is the deity of chaos versus the God of order. However, chance itself can never create the order found in the living and non-living world—as little as blindness can create sight. As the old saying goes, "what chance creates, chance destroys," because there is no purpose or direction to chance (just test it at a slot machine). In fact, it is the other way around: Chance is only intelligible in terms of the order which it lacks; a previous order must exist before any chance event can even occur. Without order, there would be no chance, for chance means being out of the order and the ordinary. If there were no order, there could be no chance, because chance needs the order of preexisting causes coming together to produce unexpected results (which is called coincidence). Besides, what seems random to us may appear to be so by lack of sufficient knowledge; "order-less" phenomena might very well turn out to be very orderly if only we had more detailed and more accurate knowledge.

Nevertheless, some evolutionists keep insisting that the world was built during evolution "from the ground up"—which they interpret as going from chaos to order, or at least from simple to complex. True enough, evolution seems to have a time arrow that moves from what is "less complex" to what is "more complex." But does this mean that the process of building more order is being steered by "disorder"? Philosophers like Daniel Bennett think it is. Bennett says that religion has it wrong—upside down—by claiming that God's creation started with order from the very beginning, whereas science places order at the *end* of an evolutionary process.[1]

I don't think he is right. Chaos cannot create order, as chaos is only intelligible in terms of the order which it lacks. I admit some scientists did develop a new interest in chaos and chaotic systems—as if these could ever falsify the existence of order. Their point is that some natural systems can only be described by non-linear mathematical equations with such complex solutions that we cannot exactly predict what the system will do in the near future. However, this isn't really chaos but only appears to be chaos; as a matter of fact, we are looking for the very order behind these seemingly chaotic phenomena, but in and of themselves, they do not falsify the existence of order. Actually, it would be very hard to search for "a law of disorder," since "disorder" and "the orderliness of a law" do not go well together.

Although science tries to explain that order comes from below, religion keeps maintaining that order comes from Above. As a matter of fact, the order we see in nature does not and cannot come from chaos, but it must come from a more fundamental order at a deeper level—which is the cosmic order of this universe governing everything that happens in the world, including the process of evolution. Compare this with the situation where shaking a jar of variously shaped candies would not create much more order, but shaking a jar with *round* candies would, because those candies have an underlying order (at a "deeper" level") that allows for a "hexagonal closest packing" structure.[2] In a similar way, the cosmic order regulates which processes are physically possible and which designs are biologically fit. Keep in mind that "law and order" are not an apriori (rooted in the way we *think* about the world), but a given (rooted in the way the world *is*). And let's not forget that statistics—the science of randomness par excellence, especially probability theory—is in itself a very orderly enterprise.

Claiming that evolution can create order out of disorder—out of the disorder of events that seem to occur "at random" or "by chance"—is only possible when we confuse randomness with disorder. Many Darwinists are caught up in this confusion. Why are they mistaken? First of all, the "simple" is in no way less orderly than the "complex." There is as much order behind the "simple" as behind the "complex."

Second, evolution cannot create order, but it must assume order. The physicist Stephen Hawking even went as far as saying that the universe created itself from nothing: "Because there is a law such as gravity, the universe can and will create itself from nothing. Spontaneous creation is the reason there is something rather than nothing, why the universe exists, why we exist."[3] I would object that this idea of spontaneous creation is sheer philosophical magic; for something to create itself, it would have to exist before it came into existence—which is absolutely impossible. As it turns out, being a good scientist doesn't necessarily make someone a good philosopher[4] —nor vice versa.

Third, evolution may seem a "disorderly" process when taken in a purely scientific context; yet, evolution could never happen without the existence of order in the background. Let's not forget that DNA-replication, reproduction, and natural selection are in essence very orderly processes.

Fourth, the study of evolution itself is certainly a very orderly enterprise—based on the assumption of order (as a precondition for confirmation and falsification), as well as on the assumption of design (as a precondition for natural selection). Because order is a philosophical assumption in science, chaos could never falsify the notion of order, given the fact that even the scientific rule of falsification is based on the assumption of some cosmic order.

So where does order come from then? Many scientists seem to think that science autonomously operates in a "vacuum," but in fact it operates within a philosophical, meta-physical framework of presuppositions. The problem is that science itself cannot be founded on scientific findings, as these presume science; its foundations must come from somewhere else. Without such pillars, science would just collapse. Just think of a game like chess: It cannot explain its own rules. Let me clarify what this entails for all sciences.

To begin with, scientists must assume, at any moment in their research, that our universe is *intelligible*. But where does this notion of intelligibility come from? It certainly does not come from science itself, of course. The word "science" is based on a Latin word for knowing, *scire*. Scientists assume that, in principle, the world can be known and taken as intelligible—otherwise there would be no need to pursue science. Besides, as to most of the sciences, this knowledge often has to be expressed in a mathematical framework in order to be considered intelligible. In other words, intelligibility is definitely not the outcome of intense and extensive scientific research; it is not intra- but extra-scientific; it's a proto-scientific notion that must come first before science can even get started.

Intelligibility can never be founded by scientific means, for it actually enables science. If you were told scientists had discovered that certain physical phenomena are not intelligible, you would, or at least should, tell them to keep searching and come up with a better hypothesis or theory—based on this fundamental philosophical assumption that says the universe is "basically" intelligible and comprehensible. Albert Einstein used to put it this way, "The most incomprehensible thing about the universe is that it is comprehensible."[5] Indeed, as far as science is concerned, that's definitely an incomprehensible thing, perhaps even a mystery. To put it briefly, science must assume intelligibility, but cannot explain it.

Something similar can be said about the notions of order and causality in all of the natural sciences. It is due to these notions that science can explain and predict—which would not be possible in a world of disorder and irregularity. Again, these notions are not scientific *out*put, but rather philosophical and meta-physical *in*put; they do not come from science but they enable science; they are the reason for trusting our reasoning. Since we may assume that like causes produce like effects, we are able to explain and predict. Do not think that science has demonstrated this experimentally, for should we find that like causes do produce *un*like effects instead, we would automatically assume that those causes must somehow be different or that some "hidden" causes must have interfered.

The same holds for the notion of order. Order is a philosophical, meta-physical presupposition, neither to be proven nor to be disproved. We can never prove the existence of order beyond any doubt, because order even holds for all cases to come as well as cases that could have happened but did not happen; so final verification is impossible. Nor can we ever disprove order, because finding any exception to the rule of order does not disprove order as such but it may falsify a specific order we had conjectured or hypothesized. In other words, the scientific rule of falsification wouldn't even work if there were no order; for it to work, it must assume order and regularity—otherwise, falsifying or refuting evidence would just be a case of irregularity in nature. The rule of falsification is based on order and cannot be falsified by disorder, so counter-evidence does allow us to falsify theories, but not the principle of falsification itself. When I hear John Stuart Mill say "It is a law that every event depends on some law,"[6] I always wonder whether he realized where that law came from. Certainly not from science!

Consequently, "law and order" are not an *a priori* (rooted in the way we think about the world, à la Kant), but a *given*, rooted in the way the world is. Einstein wrote in one of his letters, "But surely, a priori, one should expect the world to be chaotic, not to be grasped by thought in any way."[7] Einstein was philosopher enough to realize the importance of a given order, one of the main pillars of science.

And then there are the notions of design and functionality, especially so in the life sciences (as well as in the technical sciences). Camouflage is "for" the deception of predators, a heart is "for" blood circulation, eye lenses are "for" vision. However, these features can only achieve their goal of what they are "for" if they have a design that does lead to this goal—that is, if they are successful, effective, and teleological. What this means is that there is something in successful biological designs that carries them through the filter of natural selection. What then is this "something" that makes them work successfully and effectively? The answer to this philosophical question, so we found out, is that they must have followed the "rules of the cosmic design." The universe has an overall set of restraints restricting the range of possible end results.

As said before, a heart and a pump, or an eye and a camera, can and do work because of the way our universe has been "designed" and has been outfitted with teleology. Without a meta-physical cosmic design in the "background," they could not work at all. And as a consequence, natural selection can only select those biological designs that are in accordance with the cosmic design. Only by assuming *meta-physical* design, natural selection can explain specific cases of *biological* design. It is for this reason that functionality is a basic concept in the life sciences, thus making biologists search for the function of any biological feature—and if they can't find one, they will and should keep searching until they do find one. Functionality is as basic to the life sciences as causality is to all of the natural sciences.

Let me give you some examples of how dangerous it is to just trample over the quest for functionality. Not so long ago, thyroids (and even parathyroids) were completely removed from patients with goiter or thyroid cancer, until it was discovered that the side-effects were disastrous, unless patients are treated with hormones. And what to think of organs we think have lost all their functionality, such as tonsils, adenoids, or an appendix? Can they be removed without causing side-effects? Adenoids seem to be important for the immunity system of babies. The appendix may serve as a haven for useful bacteria when illness flushes those bacteria from the rest of the intestines. And what about the "junk DNA" we talked about earlier? Is it really junk, or does it have a "silent" function? Just as all scientists keep searching for causality, life scientists keep searching for functionality. It is part of their profession.

To cut a long story short, when Darwinism speaks in terms of randomness, we should be very cautious in our interpretation of this terminology. It is certainly not a denial of causality and functionality. If we take randomness to be another word for "disorder" or "Chance with a capital C," we have entered a territory where science no longer operates. That's where Darwin's legacy can easily go astray and get lost.

NOTES

1. Daniel Bennett, in *The Cambridge Companion to Atheism*, ed. Michael Martin. (New York: Cambridge University Press, 2007), chapter 8.

2. This example is analyzed in Stephen M. Barr, "Fearful Symmetries," *First Things*, October (2010).

3. Stephen Hawking and Leonard Mlodinow, *The Grand Design* (New York: Random House, Bantam Books, 2010), 180.

4. Albert Einstein was right when he said: "the man of science is a poor philosopher" (in *Out of My Later Years* (New York: Philosophical Library, 1950), 58.

5. Quoted by Antonina Vallentin, *Einstein: A Biography* (London: Weidenfeld and Nicolson, 1954), 24, quoted from Einstein's article "Physics and Reality," *Journal of the Franklin Institute*, March (1936). However, in the actual article, reprinted in *Out of My Later Years* (New York: Gramercy Publishing, 1993), the quote is: One may say "the eternal mystery of the world is its comprehensibility."

6. Mill, *A System of Logic*, chapter. V, § 1.

7. Letter to Solovine (March 30, 1952).

Chapter Eleven

No Purpose: The Good Legacy

The word "purpose" has been another cause for much of the confusion surrounding Darwinism. In general, we could say that only human beings and the Grand Designer of this universe can have purposes in mind—for only those have a mind. Therefore it shouldn't surprise us that evolution before Darwin was based on purposes the Designer of this universe had in mind. Adaptation was explained by cases of instant design based on purposes in the Designer's mind.

By its very nature, however, science feels uneasy about the concept of "purpose." To explain why this is so, we need to delve a bit into the history of science. When the *Royal Society of London for the Improvement of Natural Knowledge* started its scientific activities in 1660, it officially ratified what previous scientists had begun: It took the pagan concept of deity out of nature, because it considered nature to be a created entity—that is, created but not divine, given to us to be studied. Thus it de-deified nature, took its quasi-divine and semi-human purposes away, and made nature a non-divine, almost mechanic entity. When Copernicus entered the scene, "purpose" was taken out of astronomy; from Newton on, "purpose" has vanished from physics; and ever since Darwin, "purpose" has been taken out of biology. Science didn't experimentally prove that there are no purposes in nature, but it did start out with having "purpose" already removed from its radar screen as a premise before any experimentation could take place.

What had happened in the meantime is that the first scientists, later sanctioned by the *Royal Society of London* and other scientific societies, had wisely and clearly demarcated their area of investigation. In their charter, King Charles II assigned to the fellows of the Society the privilege of enjoying intelligence and knowledge, but with the following important stipulation "provided in matters of things philosophical, mathematical, and mechanical."[1]

That's how the "division of the estate" was executed; it was this very "partition" that led to a division of labor between the sciences and other fields of human interest. On the one hand, it gave to scientists all that could "easily" be solved by counting, calculating, and measuring, so they were allowed to investigate what's in the "Book of Nature." On the other hand, these scientists were told to keep their hands off all other domains where purposes reside—religion, education, legislation, justice, ethics, and what have you. In other words, these scientists had accepted to limit themselves to questions that could be answered by material, mechanical, physical, and mathematical means; all the rest was given away for others to handle. They were happy to relinquish those other issues, if only they were permitted to "know" what can be known by counting, calculating, and measuring. That's the domain they deemed "manageable." Ever since, all other questions and aspects have been off-limits for them and lie beyond their reach, left to the expertise of others.

By accepting such a demarcation, science bought its own territory at the expense of inclusiveness. From then on, all scientists kept profiting from this separation but, unfortunately, often forgot what it entails—limitation. Science has nothing to say about anything outside its territory; whatever is outside one's scope doesn't have to be illusionary. Alas, many scientists have come to actually reject what they neglect, but luckily there are still many left who realize that not everything that counts can be counted.

But there is another side to this story: Since anything that can be "counted" can be studied by science, "purposes" should be removed from the scientific picture as non-measurable, non-quantifiable, non-mathematical, non-mechanical, and non-physical entities. After "purpose" was taken out of astronomy and "purpose" had vanished from physics, it was Darwin's turn to remove "purpose" also from biology.

From then on, biological features were to be understood in terms of their *effects*—that is, in terms of survival problems that need to be effectively solved. As we said earlier, they serve a function; the green color of a caterpillar has a function, namely, to deceive potential predators; such is their end or goal—or in more neutral, biological terms, such is their function. But since Darwin, it is no longer their "purpose." Green caterpillars achieve a goal without having that goal as a "purpose in mind"; they were just born that way. So they have functions but not intentions. Intentions and purposes may be something in the mind of a human product-maker, but functions and designs are a feature of the product itself. They function *as if* they are steered by a purpose.

Sometimes, people who have to pop dozens of pills a day for medication wonder how those pills find their target. Or think of all hormones that flow through the blood stream and end up in their target organs. Isn't that goal-oriented behavior? Yes, in the sense that they do end up at the right target, but no, in the sense that they just bind to receptors specific for their target cells.

In other words, things like a heart, an eye, a DNA molecule, a hormone, etc. do not have a purpose, but they do have a goal—or actually an effect—that is beneficial in the "struggle for survival and reproduction." With this approach, Darwinism made evolutionary processes accessible to science and explainable in terms of natural causes and natural laws. I would say that's a very good part of its legacy.

NOTE

1. The Royal Society originated on November 28, 1660, when twelve men met to set up "a Colledge for the promoting of Physico-Mathematicall Experimentall Learning." Robert Hooke's draft of its statutes reads literally: "The Business and Design of the Royal Society is: To improve the knowledge of natural things, and all useful Arts, Manufactures, Mechanik practices, Engyries and Inventions by Experiments — (not meddling with Divinity, Metaphysics, Moralls, Politicks, Grammar, Rhetorik, or Logic)."

Chapter Twelve

No Purpose? A Not-So-Good Legacy

When scientists, including Darwin, removed "purposes" from scientific discourse, they didn't make them disappear entirely, of course; they just moved them from inside to outside the scientific domain. The fact that "purposes" don't exist in science doesn't mean they don't exist at all; they are not completely out of the picture, but they are out of the scientific picture. But you can never just reject what you neglect. As we all know, the direction of something like a billiard ball on the pool table is not only ruled by physical laws but also by human intentions and other purposes in life. Human beings have the ability to steer the laws of nature in a certain direction, according to certain purposes that they have in mind. However, once "purpose" has been eliminated from science, it can no longer be explained by science, so it is forever beyond science's reach.

In spite of the above, some Darwinists are happy to overstep their scientific boundaries by claiming that *all* processes in nature are purposeless—not only in the context of science, but as part of their worldview. Thus they claim that evolution is a purposeless, senseless, and meaningless process—which is a dogmatic ideology, nicely disguised in the garment of science, turning science into a pseudo-religion: omni-potent and omni-present.

One of its classical "high-priests" was the paleontologist and Darwinist George Gaylord Simpson from Columbia University, who ventured to proclaim from his "quasi-scientific" pulpit, that "man is the result of a purposeless and natural process that did not have him in mind."[1] That's when science has become a pseudo-religion with a monopolistic worldview. I would like to ask Simpson what the purpose is of trying to prove that there is no purpose in life. It may be very legitimate to claim there is no room for "purpose" in science, but we cannot inflate this claim into an all-comprehensive statement saying there is no purpose in this universe at all. That is an illegitimate, unwarranted move.

Isn't it startling that evolution has produced some human beings devoting their entire career to the very purpose of proving that evolution has no purpose whatsoever? Or—to paraphrase one of C. S. Lewis's discoveries—if there were no light in the universe and therefore no creatures with eyes, we should never know it was dark.[2] Similarly, if there is no purpose in the universe, how were we ever to know there is no such thing as purpose? There is no more place for purposes in scientific explanations, but that doesn't mean that purposes don't exist any longer.

Let us keep in mind that issues of sense, meaning, and purpose do not belong in scientific discourse—they cannot feature on scientific maps, and yet they do exist in the minds of map-makers. Science may be everywhere, but science is not all there is. The question as to whether evolution has a meaning, destination, or purpose takes us into the domains of philosophy and religion. Science has just no answers to these kinds of questions—no matter what some scientists proclaim, or like to proclaim. Darwinism just does not know whether evolution has a purpose or not, for the simple reason that the word "purpose" is not part of its vocabulary.

Seen from a philosophical viewpoint, there is nothing *random* in evolution. Human beings are *contingent*—that is, they could easily not have existed, since the reason for their existence can't be found within themselves—but that doesn't mean they are random creatures or mere products of a blind and purposeless fate. They may have come through a process of evolution, but ultimately they must have come here from "somewhere," or else they couldn't be here at all. In that sense, *creation* is a precondition for evolution—perhaps not as a logical but a rational "must." True, evolution didn't have humanity "in mind"—for evolution itself is mindless—but it could still be the product and purpose of the Mind of its Creator. Otherwise, we wouldn't and couldn't even be here, one could argue.

Whether you agree with this last part or not, science could never claim that it has replaced religion. I must admit many scientists think it has. Their claim is that concepts such as the Big Bang and evolution have replaced the idea that this world is a product of *creation*. Science is supposed to be an alternative and replacement for religion—and a much better one, in their view—so if evolution is in, creation must be out. They argue that, if science is right, then religion must be wrong; therefore, religion is over, or at least it's on its way out. Whenever science makes a little step forward, religion is believed to take a little step back. However, mind you, religion keeps burying its undertakers.

It's just not true that science and religion compete with each other. Why couldn't they complement each other instead? If there is creation, it must come first before the Big Bang and evolution can even get started. Only creation can create something from nothing, whereas evolution can only create something from something else. Where does that something else come from? Nothing comes from nothing! Whereas science tends to zoom in into finer and finer details, religion tends to zoom out into a larger, transcending framework. Science has a piecemeal approach, dissecting things into smaller and smaller pieces, whereas philosophy and religion try to put all the pieces back together again, so we don't lose sight of the whole.

Science may be everywhere, but it certainly is not all there is, so don't make it a know-all and cure-all. Scientists are specialists just like electricians and plumbers are specialists—they have their own expertise, nothing more and nothing less. They are specialized in applying the scientific method and in using sophisticated instruments and very advanced lab techniques. But this is not to say that scientists have some unusual, esoteric insights regarding all that's going on in the universe. Even astronomers don't deal with the entire universe, but only with its physical aspects. So let's knock science off its pedestal and put it where it belongs.

When the first scientists, later followed by the *Royal Society of London*, decided to clearly fence off their area of investigation—by limiting themselves to questions that could be answered by material, physical, and mathematical means—they created their own framework of reality, which doesn't mean that the territory they left untouched isn't "real." Yet our scientific society makes us believe that whatever is left untouched by science appears as a vacuum—and any kind of vacuum easily goes unnoticed. However, the "scientific" world is only one aspect of the "real" world; it deals only with whatever can be measured, calculated, and quantified. Limiting oneself to the material aspects of reality may be smart, but denying there is anything else is naïve, for not everything that counts can be counted.

There are many useful and truthful "maps" of this world, but each one of them is, by nature, only a surrogate for "the real thing." Each one offers us only a "one-sided" view and interpretation of reality. Science always leaves open the grand question "Why is there something rather than nothing?" Shrugging your shoulders in disregard doesn't make that question vanish. After science has explained and predicted all it can, we might think there is nothing more to wonder about, yet there is inevitably some "residue" left behind—the ground of all that exists. As the saying goes, nothing comes from nothing. That's the "void" that science permanently and necessarily leaves behind.

Science and religion do not put forward rival answers to the same question, but they offer their own answers to two different kinds of questions. Science deals with *how*-questions, whereas religion deals with *why*-questions. In other words, don't ever reply to a scientific question with a religious answer, or to a religious question with a scientific answer. Science and religion each have their own territory. As the saying goes, good fences make good neighbors; and besides, those fences act like a wall keeping "insiders" in and "outsiders" out. So let's respect those fences.

We should learn from this to never make science a pseudo-religion, nor to make religion a fake kind of science. As the late Harvard biologist Stephen J. Gould (1941–2002) once said: "the magisterium of *science* covers the empirical realm: what the Universe is made of (fact) and why it works in this way (theory). The magisterium of *religion* extends over questions of ultimate meaning and moral value. These two magisteria do not overlap."[3] They should not overlap because they are actually very different realms: Science handles the material aspects of the world, whereas religion deals with its immaterial and spiritual aspects. Demanding that these two should be merged would detract from the glory of each. They are neighbors that actually can't live without each other.

It is very much like the concept of "separation of church and state"; the state should not impose a church, nor should the church impose a state. Christianity, unlike Islam, distinguishes between political and religious authorities: "Render therefore unto Caesar…." The state should be protected from the church as much as the church should be protected from the state. Although state and church shouldn't interfere with each other, they should still acknowledge each other's role and authority. Apparently, we need them both, but not intermingled; therefore, treat them as "*both* church *and* state," not "*either* church *or* state," because they do not oppose each other but complement each other. One has what the other lacks.

Something similar holds for the relationship between religion and science. They are autonomous and independent from each other, but still need each other in order to cover what the other is missing—whereas in isolation, they are incomplete and under-performing. As Albert Einstein, a scientific authority, once said, "Science without religion is lame, religion without science is blind."[4] Or in the words of Pope John Paul II, a religious authority: "Science can purify religion from error and superstition. Religion can purify science from idolatry and false absolutes."[5] Therefore, science should be protected from religion as much as religion should be protected from science. So honor the "fences," but don't forget that, no matter on which side of the fence you happen to be, you do have a "neighbor."

What does this entail for the science of evolution? Whatever evolutionary theory may reveal about the role of accidental, random, and purposeless forces active in the evolutionary process, this leaves us only with a biological story and a biological point of view. The famous late geneticist Theodosius

Dobzhansky was wise to stress that it is in biology that nothing makes sense except when seen in the light of evolution.[6] But he was very aware of the fact that there is more to life than biology.

For now I will just urge everyone not to make biology a new religion; biology is not all-powerful or all-pervading. Some scientists may claim to have all the answers to all possible questions, including the answer that evolution is all there is and all that counts, but they actually have a very myopic view on the world by declaring everything else in life outside their narrow scope as a mere illusion. They claim that, as scientists, they know everything about the universe, based on some esoteric knowledge that everyone else supposedly lacks, but they fail to remember they are just specialists like any other specialists—specialists in doing scientific research regarding the material and physical aspects of this world, leaving everything else beyond their profession to others.

Why then is it that mainstream Darwinism is so reluctant to accept that there is still room for "purposes" left outside the scientific domain? I think the main reason is that Darwinism has often associated itself with *scientism*—the most common and popular "ism" among scientists. Scientism is the baseless ideology, the unfounded belief, the unwarranted conviction that the scientific method is the only way of finding truth. I think I have stressed enough that such a claim can only be made from *outside* the scientific realm, thus grossly overstepping the boundaries of science. How could science itself ever prove that science is the only way of finding truth? That would be like an electric generator running on its own electricity. So scientism is actually self-defeating; it violates its own rules: If it is true, it makes itself false…

Let's knock science off its pedestal and put it where it belongs. Never confuse a scientific "map" or model of the world with the world itself; maps are only a surrogate for "the real thing," so don't replace the world with one of its scientific maps. The fact that highways are missing on railroad maps, doesn't entitle us to deny their existence. The fact that "purposes" are missing on biological maps doesn't mean they don't exist at all. Indeed, some Darwinists have an ideological obsession with reducing everything to purely material causes—which makes for a very poor and empty kind of philosophy, as if science were the only way of understanding the real world, which would confuse the "scientific" world with the "real" world.[7]

To put it in a nutshell, Darwinism does not know whether evolution has a purpose or not, for the simple reason that the word "purpose" is not part of its vocabulary—which doesn't mean it cannot occur in any vocabulary. At the moment Darwinism oversteps its boundaries, it is no longer science but ideology. I consider that a bad part of Darwin's great legacy.

NOTES

1. George Gaylord Simpson, *The Meaning of Evolution* (New Haven, CT: Yale University Press, 1967), 345.

2. Lewis, *Mere Christianity* (San Francisco: HarperSanFrancisco 2001), 3rd ed., 46.

3. Stephen J. Gould, *Rocks of Ages* (New York: Ballantine Publishing Group, 1999), 207–208.

4. Einstein, *Science, Philosophy and Religion, A Symposium* (New York: The Conference on Science, Philosophy and Religion in Their Relation to the Democratic Way of Life, 1941); later also published in *Out of My Later Years* (1950).

5. In a letter to the Rev. George V. Coyne, Director of the Vatican Observatory, 1 June 1988, in Robert J. Russell, *John Paul II on Science and Religion: Reflections on the New View from Rome* (Rome: Vatican Observatory Publications, 1990).

6. Theodosius Dobzhansky, "Biology, Molecular and Organismic," *American Zoologist*, Vol. 4 (1964): 443–452.

7. Albert Einstein was right when he said: "the man of science is a poor philosopher," in *Out of My Later Years* (New York: Philosophical Library, 1950), 58.

Chapter Thirteen

No Designer: The Good Legacy

As Charles Darwin once wrote in one of his letters, "astronomers do not state that God directs the course of each comet and planet."[1] He was right on target: Comets and planets just follow laws. Why wouldn't the same hold for the process of evolution? Evolution just follows laws. If losing wing functionality (e.g., in penguins) is a regular evolutionary process, why wouldn't developing wing functionality be one too? And let's not forget that birds and planes alike must obey the laws of aerodynamics.

Early in life, Darwin did agree with William Paley and his "watchmaker analogy" that biological adaptations must be explained by instant cases of divine *design*, but then he realized that it is science's task to explain adaptation by *laws* of nature instead. If that's right, then there is no need for God to step in periodically in order to correct the things that would go "wrong" if they only followed God's laws of nature. True, the laws of nature are not necessarily identical to the laws that science has found so far, because the laws of science are merely our current knowledge of the laws of nature, and may very well be in need of revision—only time will tell. But even if laws do need to be adjusted, it's our task in science, not God's in creation.

I would say that was a very sound insight Darwin had, partly based on John Stuart Mill's rule that says "it is a law that every event depends on some law."[2] What happens by "law of nature" is something that *must* happen in this world of ours, and we speak of things covered by such laws as "natural." A "miracle," on the other hand, is not covered by any law of nature, so it must involve immediate divine intervention and is therefore said to be "supernatural." Darwinism appeals to laws and not to miracles; or to put it another way, whatever happens by supernatural intervention is not evolutionary.

The way Darwinism approaches evolution has created quite some stir, especially so in circles of so-called *Intelligent Design* (ID) theorists. They wonder how natural causes in terms of random mutation and natural selection can explain all those stunning cases of "complexity" in nature. They think that the Darwinian theory of evolution fails to account for life's subsequent complexity that they call "irreducible"—and that's where their plea for a Designer comes in as a factor additional to Darwinian causes. Francis Collins—a longtime leader of the Human Genome Project, a dear friend of religion but a skeptic regarding the ID theory—likes to describe the main intentions of ID fans in this somewhat disrespectful, yet accurate way: "When science needs divine help."[3]

More recently, these ID advocates have focused their attention on complexities as found in cell biology—in particular the intricacies of the molecular machineries that reside inside the cell. One of the classical and dear examples of ID supporters is the bacterial flagellum, which is an "outboard motor" that propels bacterial cells in various directions. Indeed, this flagellum does show quite an impressive case of complexity. The DNA expert Francis Collins describes it this way: "The structure of the flagellum, which consists of about thirty different proteins, is really quite elegant. It includes miniature versions of a base anchor, a drive shaft, and a universal joint. All of this drives a filament propeller. The whole arrangement is a nanotechnology engineering marvel. If any one of these thirty proteins is inactivated by genetic mutation, the whole apparatus will fail to work properly."[4]

It definitely looks like we have a structure here of such irreducible complexity that it seems to require a special design step through the intervention of a Designer. And indeed, ID theorists question whether a machinery of this kind of complexity could have ever arisen on the basis of mutation and natural selection alone. So they maintain that such complexity can only be produced by a Designer's intervention—an intelligent, supernatural cause—in addition to or instead of the natural processes of mutation and natural selection.

Their main argument goes along these lines: How could any of these components have evolved "by chance" unless the other twenty-nine had developed at the same time? It does seem rather evident that none of these parts could have evolved on a step-by-step basis until the entire structure had been assembled. Just think of all the other examples of complexity that may defy claims made by evolutionary theory—cases such as the human blood-clotting cascade, with its dozen or more interacting proteins. Or what should we make of the complexity of the human eye—just to mention a few examples? Natural selection won't work until the entire structure has been assembled, will it?

Darwinists have been very successful in defeating this argument. Science seems to have reached a stage where "irreducible" complexity can actually be reduced to step-by-step evolutionary processes of mutations and natural selection—without the need of invoking an additional Designer step. Major cracks are beginning to appear in the ID bastion. Let us go back to the bacterial flagellum, the poster child of ID fans. Comparison of protein sequences from various bacteria has revealed that several components of the flagellum are related to an entirely different apparatus used by some bacteria to inject toxins into other bacteria—which in itself is an ideal target for natural selection. Most likely, the elements of this structure were duplicated in the "silent" parts of DNA and then recruited for a new use, subject to natural selection again. And in turn, this process could go on for other elements. Selection just keeps working step-by-step—mutation by mutation, gene by gene—until the entire structure has been assembled.

Think of the following analogy:[5] Once the keystone is placed in an arched stone bridge, the scaffolding can be removed; from then on, removal of any part may cause the bridge to collapse, but that doesn't mean the bridge wasn't the product of a gradual construction process. So I ask you, what is left of the "irreducible complexity" of a system wherein, according to ID proponents, the removal of any of its parts is supposed to cause the system to effectively cease functioning? I would say there is less and less evidence to support the ID theory.

We may assume a similar scenario of a step-by-step selection process for the intricate cascading route of human blood-clotting, involving lots of proteins.[6] Most of these proteins turn out to be related to one another at the level of amino acid sequence, which most likely reflects ancient gene duplications. Since new copies were not essential for the original function, they could gradually evolve to take on a new function, driven by the force of mutation and natural selection. So the ID assumption that the entire cascade (or units of it) had to emerge fully functional from the very outset as a complete set of DNA sequences seems to be superfluous again.

If ID theorists raise the objection "What good is half an eye," I would reply "Better than no eye at all." This can be exemplified by the partial clotting cascades found in fish: Some of these proteins have a long history in the animal world. Isn't a partial cascade far better than none? As a matter of fact, when it needed to stop possible leaks much more quickly, our current blood clotting mechanism did evolve from a low-pressure to a high-pressure cardio-vascular system (with more proteins). Well, I would say there goes the divine intervention of a Designer as a causal and intelligent factor in biological explanations! Another "gap" has been closed, in spite of claims made by ID fans to the contrary. Besides, I wonder how complex features do have to be to get beyond the reach of natural processes. If losing wing functionality is a common evolutionary process, why wouldn't developing wing functionality be one too? I bet ID fans would counter that this latter functionality has to be "designed" first—in full and all at once.

Let me make one more remark in this context. Even if there are still cases left of so-called "irreducible complexity," which evolutionary theory cannot explain at this moment, those unexplained cases do in no way entail that the ID theory would have won the victory. If one theory loses because it cannot explain certain parts of evolution (yet), that doesn't imply the competing theory has won the argument. Let us not forget that the unexplained is not necessarily unexplainable, the unknown not unknowable, the unsolved not unsolvable; only time will tell.

In addition, there are more reasons for questioning the validity of some objections made by ID proponents. One of the main reasons is that many biological designs don't appear to be really "intelligent." Take the case of the Panda's thumb. Although they belong to the order of *Carnivora*, who all lack an opposable thumb, giant pandas came to subsist on a diet of bamboo, which requires a means to grasp shoots. So a bone in the wrist has become extended to serve as part of a grasping device—thus re-inventing the wheel by creating a "fake thumb" in addition to five fingers. The outcome is rather odd, clumsy, and inefficient, probably a disgrace to a competent Designer. But natural selection can explain why it happened this way. The quality of evolutionary designs can be as diverse as the quality of machine designs; natural selection goes for the best design among the presently available biological designs. Isn't that a reasonable explanation for the Panda's thumb? With respect to unforeseen future needs, the original design was indeed somehow shortsighted.

It is easy to bring in many more examples of *optimal* designs (the best of what's available) that are not perfect designs, let alone intelligent designs.[7] The fact that the oxygen carrying protein in human blood, hemoglobin, has a three hundred times higher affinity for carbon monoxide than for oxygen can be detrimental with even low concentrations of carbon monoxide. Let me just mention one more example of short-sighted design. In the course of their evolution, ruminants acquired very complicated stomachs (basically a set of four) and a very involved digestive routine. In this case, complexity doesn't seem a plus. Their "inner" life would have been so much simpler had they been given the right digestive enzymes from the onset. But evolution took a different path, building on what was already available from common ancestry. Is this an intelligent form of design? It seems piecemeal work to me; that's why I would call it shortsighted in comparison to future needs.

Another reason for not accepting the solution the ID theory offers us is the following. Many "designs" in nature not only look "clumsy" but indeed rudimentary, underdeveloped, undersized, stunted, or broken—again, not very "intelligent." I mentioned earlier the DNA sequence for an enzyme that makes for ascorbic acid (vitamin C) in most animals. Vitamin C is a potent anti-oxidant and is a cofactor in several vital enzymatic reactions. Many primates, including humans, have a defect in this DNA code—so it became a pseudo-gene. Consequently, these organisms must acquire vitamin C through food (which was abundant for our arboreal ancestors anyway), but they did hold on to its repetitive sequences in the "silent" section of their DNA. So what we have here is a "broken" design plus non-working repetitions of the original design. That doesn't look like a very efficient or intelligent design, no matter how we define those adjectives. Why would a Designer first design the vitamin C pathway, and then destroy it? I have never found a sensible, let alone intelligent, ID answer. Sorry, I cannot help but putting it this way: ID theorists love to impress us with examples of complex designs, while turning a blind eye to cases of rudimentary, stunted, or even broken designs. I consider that a form of highly selective perception!

That takes me to the most serious problem the ID theory faces. The ID theory turns the universe from a comprehensive whole into a defective whole that is not causally complete in itself but requires instant interventions to fill the voids God supposedly left behind in creation. ID fans tout "intelligent design" as a scientific concept, but in science, design is a metaphysical presupposition, not a physical, causal explanation. However, once taken as a meta-physical assumption, "intelligent design" can no longer be a physical cause at the same time; it can no longer be used to fill in the gaps left behind in the allegedly defective design of the universe.

As a consequence, the ID theory ends up on the unscientific side of the divide. It claims that biological complexity as a natural phenomenon is inherently beyond natural explanation, and therefore calls for an immaterial, supernatural principle. In essence, the ID theory is claiming that, in terms of regular science, the unknown is ultimately unknowable and the unsolved unsolvable; ID fans have basically given up on science ahead of time. Scientists, on the other hand, expect that all natural phenomena are explainable, since it is their basic belief that nature displays order and intelligibility, and is therefore accessible to the human intellect for further investigation. Science that oversteps its limits is arrogant, but science which does not go to its limits should be considered a failure.

Darwin's legacy teaches us not to invoke unscientific principles and theories in science.[8] In science, invoking unscientific principles is like scoring a goal during a soccer game while you are off-site; you are not really scoring a goal, since you are no longer playing soccer according to the rules. I would compliment Darwin here: When science does not go to its limits, it is a failure. Darwin made sure science does its job. That's certainly a legacy to be proud of.

NOTES

1. In correspondence with geologist Charles Lyell in 1861.

2. Mill, *A System of Logic*, chapter V, § 1.

3. Francis Collins, *The Language of God* (New York: Free Press, an imprint of Simon & Schuster, 2006), 181.

4. Collins, *The Language of God*, 185.

5. This analogy was, as far as I know, first used by A. G. Cairns-Smith, *Seven Clues to the Origin of Life* (New York: Cambridge University Press, 1986).

6. These proteins are also called "factors"; most hemophiliacs carry a mutated version of factor VIII, some of factor IX.

7. We did so already in chapter 9.

8. The history of science shows us various examples of unscientific principles: the so-called phlogiston theory, the theory of spontaneous generation, the theory that fermentation (with enzymes) would only be possible due to an immaterial "life-force" residing in a living cell, the theory of entelechy in embryology, etc. In all these cases, the theory at issue was not really overthrown by experiments but didn't meet the criteria for experiments. Louis Pasteur, for instance, did not experimentally defeat the doctrine that life can arise spontaneously, but he decided to define all air giving rise to life as contaminated with germs, thus making it a "measurable"—and therefore scientific—concept.

Chapter Fourteen

No Designer? A Not-So-Good Legacy

In the previous chapter, I tried to make the point that, in the arena of science, there is no room for unscientific principles such as *Intelligent Design*. And yet, I do agree that the ID theory is right in one respect: Science does require some form of preexisting, intelligent design of the universe, a *cosmic* design structure of order, including causality and functionality (teleology). And it is this very fact of order and functionality that makes *biological* designs functioning and intelligible—but not necessarily intelligent. The fact that all biological designs follow the "rules" of the cosmic design (in its meta-physical sense) explains why specific biological designs do function and achieve their goal. But the fact that they work successfully and effectively—which would be a *creation* issue—is altogether different from the question as to how specific cases of biological design came along—which is an *evolution* issue. Whether biological designs are successful or not is determined by the cosmic design of *creation*, but the question of how those came along is a matter of Darwinian *evolution*.

In short, biological designs are like pilot balloons, created tentatively in order to be tested against the cosmic design during the process of natural selection. If there is any intelligence in creation, it resides in the realm of cosmic design, but not directly or necessarily in the realm of biological designs. Biological designs presuppose a meta-physical design, and this meta-physical design is possibly best understood as an intelligent cosmic design of creation. The cosmic design is an intelligent design, which makes the universe intelligible and comprehensible, whereas biological designs are evolutionary products successful to different degrees, with some designs being better than others—or appearing more intelligently designed than others, if you will. If biological designs seem "intelligently designed," they do so in a derivative sense, based on the fact that the cosmic design is not only intelligible, but also endowed with order, causality, and functionality. Always keep in mind that natural selection must *assume* design but can neither *explain* nor create it.

As a consequence, we should never change natural selection into supernatural selection. Given the fact that super-natural actions do not square well with the natural sciences, we should not invoke them when natural causes will do. Among natural explanations, there is no place for supernatural explanations. All processes in nature follow physical and biological laws. It is these laws that control everything in the universe, including the evolution of life. If there is such a thing as a Big Bang, the laws of nature must have existed before the Big Bang occurred, creating the framework within which even the Big Bang had to take place. These laws are part of the intelligent cosmic design of creation, which most likely calls for an intelligent, lawgiving Creator, once we look for a rational explanation. But we shouldn't bring God back in again to steer, adjust, or redirect His own laws.[1] God doesn't need to keep the planets moving, nor does He need to keep species evolving—they just follow their God-given laws.[2]

Yet, even the famous physicist Isaac Newton did fall for this timeless temptation of having God keep a "Divine Foot" in the door, when he called upon God's active intervention to periodically reform the solar system from increasing irregularities, and to prevent the stars from falling in on each other, and perhaps even in preventing the amount of motion in the universe from decaying due to viscosity and friction. Today, we know God doesn't have to make these interventions, because science can now explain them with the proper laws (which are God's laws anyway).

It is my impression that ID fans are making the same mistake as Newton did when he invoked God's periodical interventions. Intelligent Design may be a very legitimate *meta*-physical notion—actually the only, or at least best, rational explanation for the fact that the universe is intelligible and orderly. However, it is not an *intra*-scientific, but an *extra*-scientific concept that would be considered a "foreign body" in science since it is part of the metaphysical realm.

My point is that once we lose the notion of creation, plus the trustworthy order it comes with, scientific laws become very problematic. The philosopher David Hume (1711–1776) and his followers headed down that blind alley. According to Hume (one of the first skeptic philosophers), the law of a daily sunrise is not "out there," but is merely a mental conception based on our habit of seeing the sun rise every morning.[3] Indeed, that's the trouble we are definitely in for once we disconnect science from its philosophical presuppositions, including the orderliness of creation. Hume had no reason anymore to trust his own reasoning, as he had lost sight of the fact that the laws of nature are a given, not just an acquired thought pattern; they are to be discovered, not to be invented.

If natural laws were just mental habits, it would be hard to explain why bridges built in accordance with the proper physical laws and boundary conditions stand firm whereas others collapse. Would competent engineers really have better mental habits than their inept colleagues? There has got to be more to it! G.K. Chesterton once "seriously joked" about a conspiracy of order in our world of regularity: "One elephant having a trunk was odd, but all elephants having trunks looked like a plot."[4] Well, science is in search of that plot!

In this view, the universe is a law-abiding, intelligible structure that can be best explained by the fact that it has been created by a lawful and rational God.[5] This entails that the universe is open to analysis and investigation by reason; it is an intelligible plan accessible through the light of reason. Isn't it amazing how so many scientists have been eager to adopt science, rooted in the concept of creation, but then easily lost awareness of its very foundation?

Whether they like it or not, all scientists keep living off Judeo-Christian capital. Christoph Schönborn, a theologian, put it well when he said, "Nothing is intelligible—nothing can be grasped in its essence by our intellects—without first being ordered by a creative intellect."[6] Or to quote a famous scientist, the late astrophysicist Sir James Jeans, "the universe begins to look more like a great thought than a great machine."[7] What he essentially meant was that physics is beginning to discover more and more that the universe has an intrinsic rationality, a cosmic order, which governs all that science is trying to decipher. Einstein expressed something similar: "the eternal mystery of the world is its comprehensibility."[8]

Interestingly enough, there is growing evidence that the "scientific revolution" in the fifteenth and sixteenth centuries was not really a revolution[9] but had its roots more in the Christian Middle Ages than in the world of ancient Greeks such as Aristotle and of Islamic scholars such as Avicenna.[10] The latter may have been masters in geometry and mathematics, where pure reason reigns, but this made them think they could generate scientific results deductively from passive observation, which would make those conclusions seem logically necessary. Why could it not be the other way around—that is, the opposite of what they *thought*?

Because the Judeo-Christian God is a reliable God—not confined inside the Aristotelian box, not capricious like the Olympians in ancient Greece, and not entirely beyond human comprehension like in Islam—the world depends on the laws that God willed and had laid down in creation. The only way to find out what the Creator had actually done is "interrogating" the universe by investigation and experiment. That's the approach which paved the way for experimental science. Science was invented by the first scientists to give glory to God by examining His laws of nature. Even the nuclear physicist J. Robert Oppenheimer, who was not a Christian, was ready to acknowledge this very fact: "Christianity was needed to give birth to modern science."[11]

One might in fact claim that the scientific project and its scientific method are an invention of the Catholic Church.[12] It was precisely during the high Middle Ages that the great European universities were founded. One of the first scientists, Bishop Robert Grosseteste, introduced the scientific method, including the concept of falsification, as early as the thirteenth century. The Franciscan friar Roger Bacon established concepts such as hypothesis, experimentation, and verification, so science would be free from foreign authorities and habitual bias. In other words, what some consider a period of darkness was actually the birth of the light of reason.

This takes us to the question as to what our universe and its cosmic design are populated with. What is the "fabric" of the cosmos? Mind you, this is not a scientific issue, but another philosophical, meta-physical question. The answer could run along these lines: Our universe seems to be "stuffed" with things that are marked by properties, and that are mutually connected by relationships. In other words, there appear to be three different kinds of entities in this world. First, we have things, sometimes called *substances*, such as molecules, cells, and stones. Next we have *properties*, making certain things inorganic and others organic, or certain creatures plants and others animals. Finally, we have *relationships* between things, such as the relationship a mother has to her daughter, or football players to their team. When we say "Joe is tall and single"—unlike "tall"—"single" is not a property; it is about a (missing) relationship. These distinctions seem to be rather basic and yet vital; they go as far back as the philosopher Aristotle, and can still be found in the philosophy of the late Harvard philosopher Alfred North Whitehead.[13]

This all sounds pretty evident, intuitive, and even compelling, you might say, but what does this have to do with science? The answer is: everything! Believe it or not, this discussion has important implications, even for science, and especially so for the life sciences. History tells us about a lingering debate as to how "lifeless" molecules could ever give rise to life. It's the old debate between *mechanicists* and *vitalists*. Aristotle phrased the issue this way: "The whole is more than the sum of its parts."[14] This means, to take a modern example, that a cell is more than the sum of its molecules. Well, is it, or is it not? Believe it or not, the answer depends on your philosophy!

According to mechanicists, the "whole" such as a cell or an organism is nothing more than the sum total of its parts, since only the parts are "real," whereas the rest is mere fiction. But vitalists would object to this; they claim that a cell is definitely "more" than the collection of its molecules, or that an organism is "more" than the collection of its cells—or in more general terms, that the whole is "more" than the sum total of its parts. Vitalists surely seem to have a point here, but the question is how to take this "more." Should we consider this "more" as an added, new "thing," a new substance, the way vitalists would when they introduce the vital spark of a so-called "vital principle"? Or is such a spark mere fiction?

I don't think this "more" should be considered a new "substance." Vitalists actually make the same mistake as mechanicists by only focusing on substances and properties, without paying any attention to the *relationships* and interactions between all components. I would rather state that this "more" is to be found in the very relationships between the components—that is, in the way the whole is organized. In that sense, the whole is indeed composed of parts only, and yet it is "more" than the sum total of its parts because of its organization.

Take the following analogy: A soccer team composed of the best individual players is not necessarily the best soccer team, unless these players have the relationships that make for a good team as well. You may also think of what they call synergism—when two or more agents work together to produce a result not obtainable by any of the agents independently. What would we do without those relationships? They are as vital to the cosmic design as substances and properties. If we don't take those relationships seriously, we end up in either mechanicism or vitalism.

Obviously, it is important that we acknowledge relationships in the architecture of our universe; otherwise we would completely, and perhaps unknowingly, ignore any kind of interactions between things. To take just a simple example, the properties of oxygen (O_2) and hydrogen (H_2) do not account for the properties of water (H_2O); water owes its new properties to the relationships and interactions between two atoms of hydrogen and one atom of oxygen in a water molecule. A similar story holds for the relationship between an enzyme and its cofactor; it is only when combined that they acquire a catalytic property. The enzyme alcohol *dehydrogenase*, for instance, requires the presence of zinc as a cofactor for it to work. How could we ever explain all of this if we didn't acknowledge the impact of relationships in the cosmic design?

One more case: Biologists have some idea as to which molecules are needed for life, and yet they still cannot create life in a test tube. What is required in addition is the proper organization of all the individual components—their interdependence, that is. Cells, for example, show a strong compartmentalization, which ensures that certain molecules such as enzymes are not only clustered in distinct cell compartments, but are also firmly confined to those compartments, thereby restricting reactions that might otherwise compete with one another. As a consequence, it is within the setting of larger compositions that components turn out to be capable of activities and phenomena that fail to show up in an isolated state.

Understood this way, the whole is certainly "more" than the sum total of its parts, and yet there are no extra substances such as "vital sparks" involved. As a consequence, the properties of the "whole" cannot be found in its parts alone, but arise from interrelating or interacting activities among the parts, making for some kind of synergism. This is, in a nutshell, the message of holism, organicism, and systems theory. We could summarize this as follows: $(1 + 1) > 2$.

This idea of organization acknowledges relationships as very important ontological factors. A higher level of organization is possible because of some special relationships between the constituent parts at a lower level. The higher level is somehow creating what the late physical chemist Michael Polanyi called special "boundary conditions" for the lower-level components.[15] That's why lower-level DNA can properly function and attain its "goal," because it is harnessed inside higher-level cells, and those cells, for their part, can even function more intricately when they are harnessed in organisms at a still higher level. Because of this, the "secret" of DNA is life—as much so as the "secret" of life is DNA, or perhaps even more so. The DNA world turns out to be a highly interactive world that has developed new properties based on new relationships.

People such as ID theorists who characterize the complexity of the DNA world as "irreducible" actually profess a very restricted kind of philosophy. They claim that the complexity of protein synthesis in the cell can never be explained by the mere components involved. I would fully agree to part of this claim: The complexity of the whole cannot be entirely reduced to and explained by its parts *alone*. But this doesn't mean that this complexity is irreducible. In their restricted philosophy, these people only acknowledge substances and properties, while ignoring the existence of relationships. At the very moment we take relationships into consideration as well, complexity still remains complex, but is not so "irreducible" anymore; complexity is just of a highly interactive nature.

In other words, whatever emerges during the process of evolution is not only the outcome of the substances involved and of their properties, but also of the new properties that come forth from the specific relationships between those components. The chemical element carbon (C), for instance, has such a rich potential of relationships that it harbors an unusual polymer-forming ability (at least within the temperature range commonly encountered on earth). That's why our entire organic chemistry is carbon-based, making carbon the "favorite" building block of the living world on planet earth.

Where do these substances, properties, and relationships come from? They are part of what I called earlier the *cosmic design*. All of this is part of and comes with the cosmic design of creation, and is regulated by the laws and boundary conditions of this cosmic design. Consequently, it is only because DNA follows the "syntax" of the cosmic design that it can operate as a genetic language.

Something similar occurs in the way evolution unfolds. Some people have questioned what the "selecting agent" is behind natural selection. Did Darwin replace God with the goddess of Nature—although he always vehemently denied this? It may seem so, but I would say instead that natural selection doesn't create anything new but just sorts out what is already there. In addition, I would emphasize that it is the cosmic design of creation that determines which biological designs are effective and successful.

Let me use an analogy to take this idea a bit further. A river follows a "path of least resistance" according to the "topographic design" of the landscape. In a similar way, the "stream" of evolution follows a path somehow regulated by the cosmic design behind our universe. Natural selection follows the "path of least resistance" in the landscape of the cosmic design.

The late French molecular biologist and Nobel Prize laureate Jacques Monod gave one of his books the title *Le Hasard et la Nécessité*.[16] Although I don't agree with the way he interprets this twosome, I will adapt it to my terminology and use it for the distinction between scientific randomness and philosophical necessity—that is, between the scientific *randomness* of a randomly meandering "evolutionary stream" and the philosophical *necessity* imposed by the cosmic design of the "evolutionary landscape." To put it differently, the stream of evolution actually runs in the bed of cosmic design. It is the cosmic design that channels and directs the evolutionary stream.

It may seem, initially, that Darwinism does not include concepts such as "design" and "Designer," but as we found out earlier,[17] Darwin may have thrown away Paley's "watchmaker," but certainly not his legendary watch; if he did throw something away, it was Paley's design-Designer, but not the design concept. The artifact analogy of design is as basic to Darwinism as it is to Paley's natural theology. After Darwin, the heart still existed "for" circulation; the cause of its existence may have been different, but its teleology was not.

Somehow, Darwin learned to live with the concept of *design*—which is definitely a teleological concept—because without it, he wouldn't be able to let natural selection do its work of selecting the best design for a specific "goal" or "purpose" or "end." When Darwin says that some biological designs are more beneficial than others, we should ask: in what respect more beneficial? The answer has to go along these lines: More beneficial in terms of the cosmic design and its teleology. In other words, natural selection does not explain design, but by assuming design, it can explain which specific biological designs "made" it in evolution.

It is for this reason that the built-in order of the universe could very well steer evolution in the direction of more complexity, etc. The evolutionary pathway that was followed so far was to a certain extent preordained by the intrinsic properties and relationships of the materials involved. It is because of those intrinsic properties and relationships of a created universe—that is, due to the cosmic design of creation—that we might go so far as to speculate that our world was bound by its very nature to give birth to life, arguably even to human life. As I said before, the stream of evolution actually runs in the bed of cosmic design by following the "path of least resistance." It is the cosmic design that channels and directs the evolutionary stream. But that's where science has reached its boundaries, since we have entered here the arena of philosophy, metaphysics, and religion.

The question as to whether the idea of a cosmic design requires a Designer is a related topic; it is not a scientific issue but belongs to the metaphysical domain. It boils down to the rational question as to why there is something rather than nothing and why the universe is the way it is. This universe and its cosmic design are *contingent*—that is, they could easily not have existed, since the reason for their existence can't be found within themselves. However, if there is no inherent necessity for the universe to exist, then the universe is not self-explaining and therefore must find an explanation outside itself, unless we want to give up on rationality ahead of time. Obviously, it cannot be grounded in something else that is also finite and not self-explaining (that would lead to infinite regress), so it can only derive from an unconditioned, infinite and ultimate ground, which most people would call a Creator God.

Of course, we can decide to give up on rationality by not answering such a question, or we can go for a better alternative and come up with a rational answer. I think the best rational answer would go along the following lines in the form of a few rhetorical questions:

- Could nature be intelligible if it were not created by an intelligent Creator?
- Could there be order in this world if there were no orderly Creator?
- Could there be scientific laws if there were no rational Lawgiver?
- Could there be moral laws if there were no moral Lawgiver?
- Could there be design in nature, if there were no intelligent De-signer?
- Could there be human minds, if there were no Divine Mind?

But no matter whether you agree that the answer to the above philosophical questions is "No," I think we do agree on one thing: Maintaining that *design* is not part of Darwinism and therefore has nothing to do with the study of evolution is corrupting Darwin's legacy. Although Darwin himself held a very ambiguous position in all of this, he wrote at one time, "I feel compelled to look to a First Cause having an intelligent mind in some degree analogous to that of man."[18]

Some people may think that this idea of a cosmic design comes close to the so-called *anthropic principle*. There are many versions of this principle, but it typically asserts that our universe is uniquely "fine-tuned" to give rise to life and even human life. This would entail that the universe must have somehow "known" we were coming. To paraphrase Descartes, "I think, therefore the world must be this way." It is true, if the boundary conditions of this universe were even slightly different from what they actually are, our universe would probably not exist, and neither would we. Indeed, the odds against a universe like ours appear to be enormous.

Yet it's my opinion that, worded this way, the principle is actually circular: The universe had to be this way or we would not be here to comment on it! That's rather trivial and begging the question, isn't it? No wonder objections against the *anthropic principle* have been plenty:

- If our planet had been suitable for another kind of life, it's that kind of life that would have evolved here.
- Another one: The fact that the actual fine-tuning of constants enabled the emergence of life does not entail that life was predestined to emerge.
- And then there is this one: There could be multiple, parallel universes, and ours happens to be a lucky hit.
- And what about this one: Every possible universe that can contain intelligent life will appear "fine-tuned," no matter how it came along.
- One more: It's not the physical environment of this universe that had to get fine-tuned to life, but the other way around—life had to adapt to the physical environment instead (in the same way as the soil did not develop to grow plants, nor did the nose come along to wear glasses!).
- Or finally: The principle is unscientific, since it does neither explain nor predict any physical constants. So it may be one of those unscientific principles mentioned earlier.

It seems to me that these kinds of objections are rather serious and do weaken the *anthropic principle*. Frankly, I am not very impressed by this principle—not only because it does have some weak points, but most of all because we don't really need it. I think what I have been stressing so far is located at a much deeper philosophical level. The meta-physical questions I have been raising in this book go far beyond the physical data and constants that scientists have come up with. I am more concerned with questions such as "How can electrons do anything whatsoever, let alone follow laws?" or "How can organisms do anything whatsoever, let alone survive?" or "How can biological designs be successful, let alone reproduce themselves?" or "How come this universe is comprehensible to the human mind at all?" Even if physics could ever explain why the universe is the way it is, including its physical constants, we would still be left with the meta-physical question as to where the laws come from that explain all of this.

The answer to these meta-physical questions is not to be found in some kind of anthropic *principle*, but rather in the "architecture" of the universe. The most fundamental question is as follows: Why is its cosmic structure this way? Its explanation is not to be found in itself, because the universe is a contingent entity that calls for an ultimate explanation and ground beyond itself—an orderly Creator, a rational Lawgiver, an intelligent Designer, and a moral Lawgiver. I admit this explanation is not a logical "must," but it is not mere speculation either; it is a pretty strong pointer toward something, or rather Someone, transcending the contingency of our universe. How else could we explain that this universe is orderly and intelligible? I realize the assumption of a Creator or Designer does not explain *why* the universe is this way, but it would still be the best explanation of the fact *that* the universe is this way.[19]

Let me call in again some extra help from a source far above suspicion, the mathematician Kurt Gödel. In his so-called *incompleteness theorem,* He rigorously and mathematically proved that no coherent system—not even the system of science—can be completely closed; any coherent system is essentially incomplete and needs additional "help" from outside the system.[20] Gödel even went as far as believing that we can't give a credible account of reality itself without invoking God.[21]

The only way Darwinism could reject the above reflections is when it teams up with scientism. Darwinism itself is not an "ism" in its strong sense, but sometimes it flirts with a real "ism" like scientism and thus turns itself into a real "ism" as well. But I hope I conclusively showed you earlier[22] that scientism is an unfounded belief, an unwarranted conviction, a baseless ideology. When Darwinism itself becomes an ideology, we enter the "not-so-good" parts of Darwin's legacy. I also stated that when science does not go to its limits, it is a failure. Now I have to add: When science oversteps its limits, it becomes arrogant. That's "not-so-good" a part of Darwin's legacy.

NOTES

1. This kind of God is rightly called a "god of the gaps" who often proves to be a fleeting illusion, for when the frontiers of science are being pushed back, this god would be pushed back with them as well. God should never be a "missing link" in our scientific explanations, nor should the Creator be an element or part of God's own creation.

2. We find the idea of a *harmonious universe* early in the history of science: Johannes Kepler (in his book *Harmonices Mundi*) as well as Isaac Newton called it the "harmony of the universe." Robert Boyle would call it the "grand economy of the universe" (*Micrographia*, 16). Later on, Albert Einstein would still speak of "the harmony of the Universe which we try to formulate as 'laws of nature'." (Letter to A. Chapple, February 23, 1954).

3. David Hume. *Enquiry* VII.

4. G. K. Chesterton, *Orthodoxy* (New York: John Lane, 1908), 106–107.

5. No wonder that during the age of enlightenment, deism became very popular, picturing God as the great "clockmaker." Along with it, new designations emerged such as "Supreme Being" and "Grand Architect of the Universe" (in Freemasonry), "Divine Watchmaker" (William Paley), "Orderly Harmony" (Spinoza and Einstein), or "Nature's God" (in the U.S. Declaration of Independence).

6. Christoph Schönborn, "The Designs of Science" *First Things*, January 2006.

7. James Jeans, *The Mysterious Universe* (Edinburgh: Cambridge University Press, 1930), 139.

8. See note 61.

9. Carl Sagan, for instance, in his book *Cosmos* (New York: Ballantine Books, 1980), makes it look as if nothing happened in science between 415 AD and 1543 AD.

10. See, for instance, James Hannam, *The Genesis of Science*, (Washington, DC: Regnery Publishing, 2011).

11. J. Robert Oppenheimer, "On Science and Culture," *Encounter*, October (1962).

12. Scott Locklin, "No Catholic Church, No Scientific Method", *New Oxford Review*, October (2011): 41–43.

13. Marcum, J. and G. Verschuuren, "Hemostatic Regulation and Whitehead's Philosophy of Organism," *Acta Biotheoretica*, 35 (1986): 123–133.

14. Aristotle, *Metaphysics*, 1045a10

15. Michael Polanyi, "Life's Irreducible Structure," *Science*, 160, 3834 (June 1968): 1308–1312.

16. Jacques Monod, *Essai sur la philosophie naturelle de la biologie moderne* (Paris: Editions du Seuil, 1970).

17. In chapter 10.

18. Quoted in Kenneth R. Miller, *Finding Darwin's God* (New York: HarperCollins, 2007), 287.

19. Not surprisingly, Isaac Newton had the strong conviction that God could have made the world differently if he so wished. That's why particles have certain forces and not others, or planetary orbits have certain parameters and not others (*Opticks*, Query 31).

20. Francesco Berto, *There's Something about Gödel: The Complete Guide to the Incompleteness Theorem*. (Hoboken, NJ: John Wiley and Sons, 2010).

21. See David Goldman, "The God of the Mathematicians," in *First Things*, 205 (2010): 45–50.

22. In chapter 10.

Chapter Fifteen

Gradualism: The Good Legacy

It is essential to Darwinism to search for *causal* processes that flow from mutation to variation, from variation to selection, and from selection to speciation—a sequence that should be explained by *natural* laws. In other words, Darwinism searches for small causes that may lead to larger effects—one by one, step by step, gene by gene, generation after generation, species after species. That's the core of Darwinism: Varieties are the evolutionary base for species; species are the evolutionary base for larger units such as genera—and so on upwards on the ladder of taxonomy. I would say this is a very sound scientific practice. Since science typically operates with a reductionistic, piecemeal approach, Darwinism should apply this technique to evolution as well by dissecting it into pieces—into changes on a gene-by-gene basis and selective processes on a step-by-step basis. Would it be able to also explain *speciation* by such a piecemeal approach?

This leads us to the question as to what a species is—before we can even tackle the question as to how species emerge, split, and become extinct. The concept of species is central to Darwinism: What is it that binds the members of a species together? In short, what keeps a species together and what makes it distinct from other species? These questions need to be answered before we can even begin to discuss speciation.

It is very tempting to consider a species a group of similar looking organisms. However, this is a very hazy conception: How similar is "similar," and where should we draw the line? Some people would rather adopt Wittgenstein's concept of "family resemblances."[1] Wittgenstein noted that members of a family have overlapping similarities, even if they don't have one essential feature in common.

In biology, however, the idea of overlapping similarities is still a rather ambiguous notion. Most biologists have adopted a much more precise species concept that has proven to be very successful among population biologists, the so-called biological or reproductive species concept. It was particularly developed and advocated by biologists such as Theodosius Dobzhansky from Columbia University, Francisco Ayala from the University of California at Irvine, and Ernst Mayr from Harvard University, and has been phrased in the following terms: Species are groups of interbreeding natural populations that are reproductively isolated from other such groups.[2]

According to this species concept, a species is an evolutionary unit kept together by internal gene flow and isolated from other species by intrinsic isolating mechanisms of reproduction. Its members may share some similarities, but what is more important is the fact that they are isolated from other groups by reproductive barriers. So this species concept is all about "family ties," and not so much "family resemblances." Similarities as such do not count for much, as they are merely a by-product of reproductive isolation; in theory, the members of a species could even have nothing in common but reproductive isolation.

The strength of this biological species concept is that it relates our understanding of the concept of a species to our understanding of speciation processes. It offers an explanation of what maintains and disrupts the unity of the species. To be more specific, it makes gene flow the binding force behind a species, and at the same time, it limits gene flow through isolating mechanisms.

But what should we do with those famous hybrids—crossbreeds between members of two different species (e.g., between horse and donkey)?[3] Are hybrids a problem for the biological species concept? Quite to the contrary! Hybridization may still occur after two populations of the same species have become separated and then come back into contact with each other. If their reproductive isolation had a chance to become complete, then they will have already developed into two separate, incompatible species. If their reproductive isolation is not complete yet, then further mating between the populations will still produce hybrids, but these may not be fertile. So actually, hybrids do make sense seen from an evolutionary viewpoint, but they wouldn't make sense if all species were created once and forever by the Creator. Keep in mind, before the rise of modern-time conservationists, most species didn't make it in evolution; they either became extinct or evolved into other species. It's estimated that 99.9% of all species that have ever lived no longer exist.

The outcome of this discussion is that a species is no longer defined by *properties*, based on shared characteristics, but by *relationships*, based on shared reproduction and reproductive isolation from other species. But if that is so, Darwinism could have a problem with its mechanism of natural selection, because all cases of natural selection seem to be located *within* the boundaries of the species. If it's true that a species is characterized by reproductive isolation, we should ask ourselves how on earth a new species could ever have another species as its "parent." In other words, how can a species ever change into a new species, since they are reproductively isolated?

So the real problem comes in when the theory of natural selection tries to deal with evolution *beyond* the species boundaries: How does a new species evolve? Is natural selection capable of achieving this? This issue is called *speciation*. Let's face it, no matter how long we keep applying natural selection to our enormous variety of dogs, all individuals seem to remain part of the same species and can still interbreed with other members of their species. Critics of the theory of natural selection keep hammering on this point; they stress that the fruit fly *Drosophila melanogaster*, for instance, may undergo very noticeable, dramatic mutations, but those mutant flies remain members of the same species because they can still mate with other members of their species. Wouldn't this seeming stability of a species demonstrate that natural selection is not a decisive factor in evolution?

Not really! We do have plenty of indirect indications that speciation did in fact occur in nature based on natural selection. The diversity of species in the Galapagos Islands (Darwin's "lab") has become a classical example, especially so the huge variety of finches spread out over the many islands. Interestingly enough, Darwin initially considered them just varieties and did not even mention on which island each variety occurred. How come?

Taxonomy has its so-called "splitters" and "lumpers," who distinguish themselves by splitting or rather lumping populations and species. It is evident that Darwin had to learn the observational technique of the "splitter" John Gould who told him that the specimens of mockingbirds Darwin had collected on three islands in the Galapagos were not just varieties but three distinct species. So, Darwin was not able to demonstrate that they were a product of speciation, until Gould taught him that they were real species, not just varieties. From then on, Darwin could make the case that populations on separate islands may have been isolated long enough to develop major differences through natural selection, transforming them into different species, before related populations moved in; but from then on, they could no longer interbreed with the original population (alas, we were not there when it happened). Once a population has become reproductively isolated, speciation has become a fact.

Biology discovered many more cases to corroborate the claim that speciation does occur in nature, which is a strong point in favor of neo-Darwinism. But you may wonder what its underlying mechanism is. Well, first of all, speciation may be the result of doubling-up all chromosomes, mostly so in plants, thus isolating the offspring reproductively. Usually, however, speciation starts with some form of geographical isolation. After that, due to accumulating mutations, geographical barriers lead in time to biological barriers.[4] Take, for instance, the two *Drosophila* species *pseudoobscura* and *persimilis*; they are very closely related, yet isolated from each other by habitat (*persimilis* generally lives in colder regions at higher altitudes), by the timing of their mating season (*persimilis* is more active in the morning and *pseudoobscura* at night), by behavior during mating (the females of both species prefer the males of their respective species), and by sterility of hybrid males. Even if the original physical barrier no longer exists, the new species may stay reproductively isolated due to its acquired biological barriers. Once geographical isolation has created reproductive isolation, speciation has become a fact.

But there is more. Not all geographical isolation is based on physical barriers such as mountains, deserts, or water.[5] There are also "barriers" of a different nature—such as the mere physical distance between some members of a population. In such cases, the population is still continuous, but nonetheless, its members may not mate randomly, because they are more likely to mate with their geographic neighbors than with individuals at a greater distance.[6] How could such a situation lead to speciation?

Let us consider the case where the species forms a ring of neighboring populations who can interbreed with adjacent populations, but there may be at least two "end" populations in the series that are too distantly related to interbreed. An example of this situation can be found in California, where the *Ensatina* salamander forms a horseshoe shape of populations in the mountains surrounding the Californian Central Valley. Although interbreeding can happen between each of the 19 populations around the horseshoe, the salamanders on the western end of the horseshoe cannot interbreed with the salamanders on the eastern end. The problem, then, is whether to classify the whole ring as a single species (in spite of the fact that not all individuals can interbreed) or to consider each population as a separate species (in spite of the fact that there is still interbreeding with nearby populations). This much is clear: If enough of the connecting populations within the ring perish to sever the breeding connection, the remaining members would have become two distinct species. It shouldn't surprise us then that, in an evolutionary context, a clear-cut species concept inevitably becomes a bit fluid.

So the bottom line is this: Speciation is accumulated variation as a result of natural selection and is enforced by geographical or other forms of isolation.[7] However, there is a major problem left. If speciation is considered by some as a potential challenge to Darwinism, what to make of those differences between species that are so major and drastic that we are not only crossing borders between species but even between classes and orders in biological taxonomy? Just think of amphibians: They grow up in water (like fish do), move to land (like reptiles do), and come back to water to lay their eggs (like fish again). And then we have those reptiles; most of them lay eggs, but some species simply retain the eggs until just before hatching, others provide all nutrients via a structure similar to the mammalian placenta. Or think of the still existing intermediate form of an animal called *Platypus* which lays eggs developed in a "mammalian" uterus.

Indeed, in terms of neo-Darwinism, it is a very hard stretch to explain the transition from fish to reptiles (which is definitely more than changing swim bladders into lungs, or fins into legs), from reptiles to birds (which is, again, more than changing legs into wings), and so on. The gradualism of neo-Darwinism seems to fail us here (although it does allow for variations in evolutionary speed).

Perhaps the still controversial idea of *saltationism* might help us here to not only think in terms of small changes but also large leaps—that is, some kind of wide-ranging mutations. At this moment, it is hard to demonstrate such large-scale mutations. At Darwin's time, there was no field of genetics yet to back up his claim of *small* modifications; nowadays, genetics does support Darwin's claim but is not far enough yet to open the black box of *sweeping* genetic changes.

Yet, the idea of saltationism may not be too far-fetched, given the fact that genetic programs not only contain "blueprint" genes but also instructions for processing the "blueprint." The Y-chromosome of males, for instance, hardly carries any "blueprint" genes for specific characteristics, yet this chromosome has quite an impact on the process of sexual differentiation, as we know from rare cases when it's missing.[8] Clearly, we are still in largely uncharted territory here—with the magnitude of the unknown being unknown as well!

Although Darwin kept stressing that nature doesn't make big leaps[9]—there is his notorious "blind spot" for discontinuity again—nature is actually full of giant leaps. Sometimes small genetic changes may have far-reaching effects. The extensive metamorphosis from caterpillar to butterfly, for instance, is often controlled by one single gene. Another sweeping effect of a simple mutation can be found in humans with the gene called FOXP2, discovered in a single family in England where members had severe difficulty in speaking. As it turned out, these affected members had a single letter of the DNA code "misspelled" in the FOXP2 gene on chromosome 7. This gene has been remarkably stable in almost all mammals, until two significant changes have occurred some hundred thousand years ago in human family tree. Not only do the human and chimp versions of FOXP2 look different but they function differently as well. The FOXP2 gene switches other genes on and off, thus driving target genes to behave differently in both species. It could very well be that all these seemingly minor genetic changes had a major impact on the development of language.

Another striking example of evolution beyond the species level can be found on the island of Madagascar. For a long time, Madagascar has been in the process of drifting farther away from the east coast of Africa. During this period, a peculiar group of small mammals, *tenrecs*, could prosper on the island by lack of competing invaders of the mammalian type; they remained rather primitive (with poor vision, poor temperature regulation, a primitive excretory system like in reptiles and birds, etc.), because they had nobody but one another to contend with. Yet they were able to exploit many niches still left unoccupied (but filled elsewhere by more advanced creatures).

During this evolutionary process, *tenrecs* could preserve their simple body plan but were able to develop several specialized characteristics; as a result, they are widely diverse and have evolved into groups that resemble animals on the mainland as diverse as hedgehogs, shrews, opossums, mice, porcupines, and even otters—although they are not closely related to any of these groups, which indicates they must have gone through their own, parallel evolution. Nowadays, we distinguish 30 species of *tenrecs* classified into 10 genera and 4 subfamilies. The differences between these groups are quite dramatic, and yet we have no explanation other than that they must have had a common ancestor—some insectivoran-grade, primitive mammal from Africa—from which they must all have inherited their simple body plan.

Saltationalism may still be a very controversial topic, but think of it along the following lines. Adaptations may be variations of a common structure—say, the external skeleton of insects versus the internal skeleton of vertebrates—but it is doubtful whether the structure itself also is an adaptation. Consider the following question: If there are genes for eye-color, color-blindness, etc., would the sum of all of these be enough for having eyes? The best answer to this question is probably that a genotype is more than a genome. A *genome* is just a cluster of genes, but a *genotype* is the blueprint of a genome plus the instructions for processing the blueprint. In other words, a gene does what the program dictates—not the other way around: The program doesn't do what a gene dictates. Unfortunately, the "genetics" of the genotype is currently not as well developed as the genetics of the genome; some call this epigenetics.

If the above view is right, then there are not only reproductive borders (which separate species from each other) but also structural borders (which separate higher taxonomical units or taxa from each other). As a result, we would be dealing with thresholds at different levels. The species border is low enough to permit adaptation and high enough to prevent reproduction. On the other hand, higher taxa would have higher thresholds, making transitions punctuated rather than gradual. Is this in violation of Darwinian orthodoxy? It depends on how you look at it. In general, I would say there are different degrees of trans-specific steps, depending on which structural border has been crossed. But no matter what the outcome is, Darwin's legacy could very well remain standing.

NOTES

1. Ludwig Wittgenstein, *Philosophical Investigations* (Malden, MA: Wiley-Blackwell, 1991), 66–71.

2. Ernst Mayr, *One Long Argument—Charles Darwin and the Genesis of Modern Evolutionary Thought* (Cambridge, MA: Harvard University Press, 1991), 29.

3. Hybrids between a horse (with 32 pairs of chromosomes) and a donkey (with 31 pairs) are called either a mule (if the father is a donkey or "jack") or a hinny (if the mother is a donkey or "jenny"). Most hybrids are infertile because the chromosomes have to match up properly to create viable embryos. However, female hybrids (both mule and hinny) can sometimes still produce offspring, but all male hybrids are infertile (probably because they miss a second X-chromosome).

4. Examples of biological barriers are differences in habitats, breeding times, mating rituals, sexual organ structures, infertility of hybrids, and incompatibility between sperm cells and egg cells.

5. This phenomenon is called *allopatric* speciation.

6. This phenomenon is called *parapatric* speciation.

7. There is even experimental evidence that speciation does occur. In the late 1980s, William Rice and G.W. Salt bred fruit flies, *Drosophila melanogaster*, using a maze with different choices of habitat such as light/dark and wet/dry. Each generation was placed into the maze, and the groups of flies which came out of two of the eight exits were kept apart to breed with each other in their respective groups. After thirty-five generations, the two groups and their offspring had become reproductively isolated because of their strong habitat preferences: They mated only within the areas they preferred, and so did not mate with flies that preferred the other areas. The habitat split had actually become a reproductive split. And then there was Diane Dodd who was able to show how mating preferences can develop into reproductive isolation in *Drosophila pseudoobscura* fruit flies in as little as eight generations using different food types, starch and maltose. Dodd's experiment has been easy for many others to replicate, including with other kinds of fruit flies and foods.

8. People with Turner's syndrome, for instance, have only one X-chromosome, and no other X- or Y-chromosome, which is rendered as XO, where O stands for a missing chromosome.

9. The idea that nature makes no leaps ("*Natura non facit saltus*") does not come from Darwin literally, but can be found in Asa Gray's review of Darwin's *On the Origin of Species*. In *Atlantic Monthly*. 6 (July–August): 109–116, 229–239, Gray says: "[…] do not suppose that Nature in the organic world makes no distinct steps, but only short and serial steps, not infinitely fine gradations, but no long leaps, or few of them."

Chapter Sixteen

Gradualism? A Not-So-Good Legacy

Darwin's legacy of gradualism becomes extremely pressing when Darwinism has to deal with the origin of humanity. In essence, Darwinism is an ideology of equality: It assumes there are no real differences in the living world—no high and low, no discontinuities but just ongoing and gradual continuity. Darwinism is basically egalitarian in outlook and preaches equality, declaring that all organisms evolved equal as a result of small changes—in accordance with Darwin's famous admonition that the difference between animals and man is one of degree, not kind.[1]

I would say that, in terms of genetics, anatomy, and physiology, Darwin might be perfectly right. But what makes us believe that humans are mere animals? In line with the doctrine of Darwinism—which goes a bit farther than the basics of Darwinism justify—humans ought to be degraded to glorified animals, or reversed, animals ought to be inflated to pre-humans, humans-in-the-making so to speak. Thinking differently would be considered elitist in terms of this ideology.

So it is due to Darwin's gradualism that biologists have been searching for pre-human features in the animal world, especially in the world of higher primates. And they did search hard, coming from different angles. One of those is the angle of self-expression, so famously human. They did so, for instance, by letting apes create paintings; yet the colorful mess apes produced under pressure is far from any representational kind of art that we know of among humans. Or they like to point at the beautiful tunes songbirds can generate, but those are mere repetitions of the same, old tune—certainly not music.

They have also attempted to make apes use some kind of language—for apes don't do so on their own. However, their simple "sentences" turn out to lack any kind of grammar—something like "me Tarzan, you Jane." But we all know that the difference between "Help me accept" and "Accept my help," or between "Your trouble" and "You're trouble," or between "Be well, so you can do good" and "Be good, so you can do well" is a matter of ... well, grammar. Besides, apes have to be forced to "say something," whereas children do develop this capacity naturally (otherwise we would call upon a psychologist for help). Apparently, *Homo sapiens* is also a *Homo loquens*. The "grand old man" of behavioral studies on primates, David Premack, came slowly but eventually to the conclusion that, in his very own words, "human language is an embarrassment for evolutionary theory."[2]

Apparently, there is more to human self-expression than Darwinism seems to acknowledge. The human mind has this astounding capacity of expressing itself in various forms of art—in literature, in paintings, in music, in sports. We find these expressions as far back as the early history of humankind. Of all animals in prehistory, only humans left stones behind with inscriptions; only humans dress themselves up; etc. And then we have this astonishing art in prehistoric caves. Understandably, some Darwinists considered such paintings fake for a while, because they could not accept such a high level of self-expression that early in human history based on their ideology of gradualness. My simple response to them would be: Even a cave man can do it! Some of the first Cro-Magnon sites, dating from well over 30,000 years ago, have even yielded evidence for music: multi-holed bone flutes capable of producing a remarkable complexity of sound.

In addition, we see for the first time in prehistory evidence of elaborate burials; the amount of effort put into the aesthetic productions found in prehistoric graves suggests that decoration and art were an integral part of the lives and societies of the people who made them. Besides, burial of the dead with grave goods indicates a belief in an after-life, for the goods are there because they are considered useful to the deceased in their future lives.

How far had these people come, given their animal ancestry! How different this is from female apes that keep carrying their dead newborn around for quite a while, without having any idea of what is going on, until they finally give up and drop the dead remains. In Guinea, W. Africa, for instance, chimp mothers have been seen in nature carrying and grooming their offspring's lifeless bodies for up to 68 days. By the time the corpses were finally abandoned, the bodies had mummified and developed an intense smell of decay. Believe it or not, but some Darwinists give this observation their own twist: These must be 68 days of actually mourning the dead! All I can say in reply is that only human beings could come up with such an explanation. I would not call them "68 days of mourning" but "68 days of ignorance." Besides, I would like to question Darwinists where the burial and the grave are. Apparently, only the "finite" human mind is able to catch a glimpse of the Infinite.[3]

How come animals lack any form of self-expression? The reason is rather simple: They have no real "self." Although monkeys are actually able to use mirror images to locate things and even other animals in real space, they have been given months, not to mention years, in which to recognize themselves in mirrors, yet they continue to either neglect their mirror image or treat it as a potential aggressor. On the other hand, I must admit that chimpanzees and orangutans (unlike gorillas) do show some capacity of recognizing themselves in a mirror (as do elephants and dolphins). After marking the animal with a visible as well as a not-directly visible dye spot, scientists have been testing if and how the animal reacts to the visible dye located on its own body while ignoring the control dye.

This so-called mirror test was devised in 1970 by the psychologist Gordon G. Gallup in order to detect self-awareness by determining whether an animal can recognize its own reflection in a mirror as an image of itself. Although the mirror test keeps ranking high on the Darwinists' agenda, it doesn't entitle them to equate self-recognition to self-reflection, let alone self-awareness. Why not?

First of all, mirror self-recognition is at least partially a learned behavior. When first exposed to a mirror, most apes react to their reflection as they would to another member of the same species; but eventually they may learn to recognize their mirror image as a reflection of their own body. Second, one could argue that passing the mirror test is merely evidence of superior cognitive abilities, possibly based more on face-recognition than a sense of "self." Dogs, for instance, perform poorly on the mirror test, but they do seem to be able to recognize their own scent. Third, animals are very well able to classify, distinguish, and recognize faces and scents; so why would they not be able to classify a face or scent as coming with one particular body, their own. After all, they can associate pain with their own body. Fourth, the mirror test may show us that some animals can recognize their own body; however, being aware of their own body is not necessarily self-awareness. Animals that regularly drink from water surfaces should be used to seeing a reflection of their own bodies in the water anyway. But do they have a sense of "self," a sense of past and future, knowing that they exist in a particular time and place? I doubt, for mirrors mirror a body, not a "self." Animals won't just make a trip to the water to see them-selves in the "mirror," whereas humans tend to frequent their mirrors. And finally, the mirror test seems to tell us much more about ourselves than about the animal world. When we see apes looking into a mirror and recognizing "themselves," we hope (or fear) to see ourselves in them. To put it briefly, animals may recognize themselves, but not "their selves."

Apparently, it is not really the brain that sets us apart, not even our DNA, for that's not very different either. Unlike *bodily* coordination, *mental* orchestration is not a matter of anatomy, physiology, genes, or brains. The brain is a neuronal model of the mind; the brain is just a material vehicle for an immaterial mind—or put differently, the mind uses the brain as a programmer uses a computer. It is this immaterial mind that gives human beings such an endless potential that they can even picture themselves in make-believe situations—never seen, never experienced before. The sky is the limit. Just ask yourself the question why orangutans are still swinging from trees, while humans are able to write books, ride bikes, drive cars, and fly planes—something apes can't even think of doing. Humans use falcons, hawks, and hounds to help them hunt, but I am still waiting for them using *us* to catch their prey.

Isn't it amazing how the "Not-So-Good" Legacy of egalitarian Darwinism has produced the slogan of "Apes 'R' Us," as if nothing new emerged when humanity originated—and yet, a lot seemed to be ground-breaking when humanity came along, but the differences may very well be beyond the reach of science. Once you have been trained enough to watch only for gradualness and continuity in evolution, you may not notice anymore how much the human race differs from the animal kingdom, because the physical differences are so major indeed that they make the non-physical differences appear minor or even vanish from sight.

In addition to the issue of self-expression, Darwinists have also tried to tackle the problem of human intelligence and rationality. When it comes to intelligence, humans do not seem to be exceptional creatures. I realize intelligence is a rather complex entity, as it has many faces—social, practical, formal, spatial intelligence, and so on. Yet, many animals show some form of intelligence in their behavior, because intelligence is a brain feature and as such an important tool in survival. Animals do show various forms of intelligence: We find spatial intelligence in pigeons and bats, social intelligence in wolves and monkeys, formal intelligence in apes and dolphins, practical intelligence in rats and ravens, to name just a few.

But let's not go overboard. Almost a century ago, Edward Thorndike, the pioneer of American experimental psychology, complained bitterly about all the books of his day (published after Darwin) stating that they did not provide "a psychology, but rather a eulogy of animals. They have all been about animal intelligence, never about animal stupidity."[4] He was right, there's certainly a lot of "stupidity" in the animal world when it comes to formal intelligence. Just think of the following cases:

- The cat that was placed in a closed box and then discovered after many trials how to pull a cord in order to open an escape hole, still keeps pulling the cord before crawling out, even when the escape hole is already open.
- Or the cat that had finally learned to escape by pulling a cord, but when subsequently placed in the same box without the cord, goes through the motions of pawing the air where the cord had been before.

- The elephants that had learned to lift a lid to retrieve food from a bucket, didn't "get it" when the lid was placed alongside the bucket while the food was simultaneously placed inside the bucket; all trained elephants continued to toss the lid before retrieving the reward, raising the possibility that they have no understanding of this simple causal relationship.
- The chimpanzee that doesn't know how to remove an obstacle in order to reach food.
- The chimp that has learned to invoke help from a trainer but does not know which trainer to call upon if one of them has a bucket over his head while the other has not.
- The chimp that cannot learn to secure a larger reward by pointing to a smaller one; the immediate urge for food just overrules any intelligent alternatives.
- And my favorite one: Some tribes in Africa make a split in a coconut and fill it with candy as a way to catch monkeys who put their hand inside the coconut to grab the candy, but can't get rid of the coconut because they keep their fists tightly closed around the candy.

As it turns out, animals may have some intelligence, depending on their brain structure, but they show no clear signs of abstract reasoning or having reasons for their "thoughts" (if they have any); they are "moved" by motives, drives, instincts, and training, but not by reasons or concepts. Reasoning is pondering realities beyond that which we experience through our senses. Animals, on the other hand, seem to live their lives entirely in the present, without having any thoughts about the past or the future. If animals have a pedigree, it's thanks to their owners; if they have birthdays, wish lists, appointments, or schedules, it's because their owners create those; and if they have graves, those were dug by their owners as well.

Have you ever seen animals dressing themselves or training other animals or keeping them as their pets? What a disparity! Only humans are able to keep pets and train them. Only humans can complain that they felt treated like animals. Only humans are conscious of time; they can study the past, recognize the present, and anticipate the future; they even desire to transcend time, thinking about living forever. Only humans know God. Only humans wonder "what caused or will cause what and why?" Only human beings have inquisitive minds asking questions such as "Where do we come from?" They are always in search of some kind of worldview or explanation of life—which certainly goes far beyond their need for food. You may consider philosophy a pretty sophisticated enterprise, but there isn't probably one person who doesn't philosophize; each one of us is destined to start philosophizing at some point in life; reading this book may be one of those moments.

Rationality, based on reasons and thoughts, is closely connected to using mental *concepts*. Let me try to make my point more compelling by expanding on what it is that makes mental concepts so unique. Human beings have a mental concept of what a physical thing is and can do; it is through mental concepts that they can transform "things" of the world into "objects" of knowledge.

Not so with animals. If we train a dog to associate a command like "The boss!" with its real boss, then that dog has been conditioned to respond to the command "The boss!" by looking for the real boss. That sound has become a *signal* for the "real thing." Signals depend on the actual presence of the "real thing." The dog has a physical image of its boss, but it has no mental concept of what a boss is like. Human beings certainly share this feature with animals; one employee yelling "The boss!" is actually signaling all others present to start looking extremely busy. That's what signals do. Monkeys, for instance, have different signals for different types of enemies, but they only use them when such an enemy is around; they don't just "chat" about their enemies.

But what is really unique to humans is the fact that they can also use the phrase "The boss" not as a signal but as a *symbol* for a "mental object" of a boss. As a matter of fact, when they use that phrase, the real boss may not be around, or they may only do so when the real boss is in fact physically absent. Instead, they use their "mental object" of a boss to debate, for example, why their real boss doesn't qualify as a boss, or which real boss they would prefer instead. In other words, they are actually comparing the "real thing" (perhaps far away) with a "mental object."

Mental concepts transform "things" of the world into "objects" of knowledge, thus enabling humans to see with their "mental eyes" what no physical eyes could ever see before. Animals live in a world of events, but humans also inhabit a world of facts—which are mental interpretations of those events. Without concepts, there wouldn't be any facts, for facts are the objects of our thoughts. Imagine where science would be without facts and without the use of hypotheses based on mental concepts. The answer is, "Nowhere!"

The difference between *signals*, based on physical connections, and *symbols*, based on mental concepts, also explains that animals, depending on their intelligence, occasionally do show some capability of seeing "physical" connections between things. Because of this, animals are able to perform trial-and-error activities. Chimps, for instance, have been able to "see" that a stick can help them reach food by making such a "physical" connection when they see a stick near food; but if the food and the stick are kept far apart, they cannot make the connection—even if they were able to do so before. Apparently, the stick is not a mental concept but rather a physical signal that can invoke an intelligent reaction; a chimp has a physical image of a stick in relationship to food, but it has no mental concept of what a stick is and can do in various situations. Why not? Animals are fully enveloped in their surroundings, but they are not *subjects* capable of creating mental *objects* transcending the immediate situation; animals are caught up in a network of physical connections, unable to transcend the current situation with the mental power of abstract concepts and reasoning. That explains why intelligence differs fundamentally from rationality.

Humans, on the other hand, can do trial-and-error operations in their minds first—before they give it a real try. The human mind of a tool maker, for example, is able to create tools according to some mental concept of a tool object with which they constantly compare the real product they are working on, thus anticipating the future by means of abstract rational conceptions residing in their minds. The human mind is able to conceive of any object's properties and relationships in various situations—and this is exactly the point where the rationality of our intellect differs from intelligence of our brain.

All of this has some important implications: Intelligence seems to be a *brain* issue, but rationality is more of a *mind* issue. Some Darwinists may have brainwashed us enough to believe that intellect and intelligence are the same, and that the mind is just a brain issue—in defiance of the fact that, even if we would ever understand the human brain, we will never fully understand the human mind, as it requires a mind to understand the brain and it requires a human subject to study any objects. It is obvious, from a philosophical point of view, that the knowing subject must be more than the known object, just like the projector must be more than the projected images. So what is the brain? It is the *known* object! And what then is the *knowing* subject? Not the brain but the mind! It requires a subject with a mind to make sense of pictures or words, so they become objects of knowledge based on mental concepts. In other words, without "I" as a subject, those items are meaningless and don't make any sense.

So I think we should come to an important conclusion: Even when they study the human brain as an object of science, scientists also need the human mind as the subject of science—for without the human mind, with its intellect and rationality, there would be no science at all. One would need a mind before one can study the brain! That's why biology can never fully comprehend the human mind because biology itself depends on the working of the human mind. How could the brain ever study itself? That would be like the magic of a projector projecting itself or a copy machine copying itself.

It always amazes me how some Darwinists like to downgrade the human mind while touting their own minds. If thoughts were merely the product of bodily and other natural actions, all thoughts would be equivalent, and we would have no way of telling true from false, or knowledge from error. That takes us back to the question Darwin himself posed long ago: Can science done by a mind that science has found to be the result of natural selection be trusted? Again, the answer is no—unless we are ready to derive the rationality of human beings from a higher source.

It is this rationality that is entirely part of the mind. Some like to call the human mind the human soul; that's fine with me, as long as we realize that the mind is the intellectual part of the soul. I would say the mind is the soul's eye, its light—or put differently, the mind is the power of the soul by which we know truth. We have definitely entered meta-physical territory here.

Is the above answer acceptable to Darwinism? It is not acceptable as a scientific answer, of course, because it is located outside the scientific domain. But it could be acceptable as a meta-physical answer, unless Darwinism pairs up with scientism again and rejects any kind of meta-physics beyond the physical realm. Some Darwinists may feel parachuted on an almost uncharted terrain outside the territory they are familiar with. True, by its very nature, Darwinism must focus on the physical dimension of human evolution, but it shouldn't lose sight of the fact that there is also a non-physical dimension to it.[5]

Let's at least acknowledge that Darwin did confuse us—and was probably confused himself—when he spoke of "the *mind* of man, which has, as I fully believe, been developed from a *mind* as low as that possessed by the lowest animal."[6] Had he used the word *brain* instead, he could have made a good case. But the mind is more than the brain, and the intellect is more than a matter of intelligence. What's in a word? Sometimes, a world of differences ...

NOTES

1. In his introduction to a book chapter of *The Descent of Man* (1871/1896 p. 66), Darwin stated: "My object in this chapter is to show that there is no fundamental difference between man and the higher mammals in their mental faculties."

2. David Premack, "'Gavagai!' or the Future History of the Animal Language Controversy," *Cognition*, 19 (1985): 281–282.

3. See also a more technical assessment: Derek C. Penn et al., "Darwin's mistake: Explaining the discontinuity between human and nonhuman minds," *Behavioral and Brain Sciences,* 31 (2008), 109–178.

4. Edward Thorndike, *Animal* Intelligence (New York: Macmillan, 1911), 22.

5. Interestingly enough, it was Isaac Newton's conviction that the mechanical design of material beings prompts the mind to assume a non-mechanical, spiritual cause (*Opticks*, Query 28).

6. Charles Darwin, *The Autobiography of Charles Darwin* (Cambridge, U.K.: Icon Books, 2003), 149. Or from Charles Darwin, *Autobiography* (1876), in *The Life and Letters of Charles Darwin*, ed. Francis Darwin, Vol. 1 (London: John Murray, 1888), 307.

Chapter Seventeen

No Thresholds: The Good Legacy

Even if we accept that Darwinism has an adequate mechanism for speciation, we are still left with at least three evolutionary hurdles that seem to be hard to explain within a Darwinian framework: the origin of life, the origin of sexuality, and the origin of humanity. We will discuss here how Darwinism has tried to explain all three steps within its framework. The first two steps require some technical terminology that you may want to skip; that's why they are printed in a smaller font.

The origin of life took place some 4.5 billion years ago. If all life emerges from "parents," how could life itself emerge? Although there was no reproduction before life existed, Darwinism assumes that we can apply something like "natural selection" to the very origin of life too. I will briefly explain the theory of the Belgian biochemist and Nobel laureate Christian De Duve. I apologize for this technical in-depth section. Skip it if you wish. But I need to discuss this issue here, because it is an important topic in anti-Darwinian rhetoric.

According to De Duve's reconstruction,[1] emerging life went through four main successive stages—the primeval pre-biotic world, the thio-ester world, the RNA world, and the DNA world. His version of the script differs from the current favorite mainly by insertion of the thio-ester world, which De Duve considers essential because he cannot accept the view of an RNA world arising through purely random chemistry. We will only discuss some central points.

Stage 1 was characterized by an abundant source of energy (ultraviolet light) and an abundant source of electrons (from Fe^{2+} ions, which are essential on a planet that was still lacking molecular hydrogen). Thanks to this combination, the atmosphere became enriched with molecules such as H_2, CH_4, NH_3, HCN, H_2O, and H_2S. Through synthetic reactions sparked by electric discharges and other physical forms of energy, these were synthesized into small organic molecules such as carboxylic acids, including amino acids, hydroxyl acids, and thiols (based on H_2S). These substances accumulated in oceans and lakes as a "prebiotic broth." In the laboratory, such synthetic reactions were repeatedly confirmed.

Stage 2, the thio-ester world, is deemed necessary because the "prebiotic broth" would have remained "sterile" without additional help of both catalytic and energetic nature. This step was accomplished by thio-esters arising from the condensation of carboxylic acids with thiols. Thio-esters are capable of linking amino acids together into proteins, thanks to energy derived from the consumed thio-ester bonds. Although perhaps random and undirected, the polymerization process had a highly selective yield, because many combinations were too unstable to reach steady-state levels. The small subset of polymers that "survived" included a number of crude catalysts foreshadowing present-day enzymes. Is this real natural selection? Not in a strict sense, of course; call it "molecular selection" if you wish.

Why are these "enzymes" so important? Well, a modern cell can only develop structural proteins with the help of DNA, RNA, and enzymatic proteins. The making of proteins is currently based on DNA and RNA, but DNA and RNA can only achieve this by using enzymatic proteins. So we have a chicken-or-egg problem here: no DNA and RNA without proteins, and no proteins without DNA and RNA. But the thio-ester world would solve this problem. In addition, scientists have discovered some special RNA molecules (*ribozymes*) which can also function as enzymes.

Stage 3, the RNA world, developed a small set of proto-tRNAs. Molecules that possessed a site capable of specifically binding a given kind of amino acids were particularly favored by "molecular selection." Reciprocally, the amino acids that were recognized by proto-tRNAs were themselves selected to be used in a metabolism that depended on RNA binding. The central process of protein synthesis was thus born—but essentially with the help of thio-esters.

Stage 4, the DNA world, went through a relatively simple transition which required only a set of enzymes for reverse transcribing of DNA back into RNA. Except for RNA viruses, believed to be vestiges of the RNA world, RNA *replicase* and *reverse transcriptase* enzymes have largely disappeared from the living world. But once the transition was made, the scene was set for the first cells—some 3.5 billion years ago. The first cells resembled present-day bacteria in their structure and their main metabolic attributes (*pro-karyotes*). They occupied hot and acidic sulfurous waters; they were anaerobic in their metabolism, for there was no free oxygen yet. All the forms of life that compose the biosphere today are believed to have arisen from the further evolution of this primeval cell population.

There is at least one more step left: How did prokaryotic cells without mitochondria (the "batteries" of a cell) develop into eukaryotic cells that do possess mitochondria? There is more and more evidence that cell organelles of eukaryotic cells may have originated as free-living bacteria that were incorporated inside a cell.[2] Interestingly enough, mitochondria still contain their own DNA, different from that of the cell nucleus. These so-called endo-symbionts could even have transferred some of their own DNA to the host cell's nucleus, since we know that transfer of DNA does occur between bacteria species. Let's leave it at that.

By way of conclusion, let me stress that biologists take these kinds of processes to be random in the sense we discussed before.[3] Yet, the pathway followed was almost entirely preordained by the intrinsic properties and relationships of the materials involved, given a certain kind of environment or succession of environmental conditions. Of course, there were statistical fluctuations involved, but the statistical game was not played at the level of individual amino acids or nucleotides, but at the level of modular assemblies—some assemblies being fragile, others rather stable due to the intrinsic properties and relations of their composing parts. It is because of those intrinsic properties that we might even claim that our world was bound by its very nature to give birth to life, and perhaps even to human life. But that's where science has reached its boundaries. Let's respect those.

The second "enigmatic" step is the origin of *sexuality*. The dichotomy of male and female seems to be standard in the plant and animal world. But let this statement not fool you. Some plants harbor both sexes in the same flower; some animals do something similar (hermaphrodites) or can even change their sex. And yet, there seems to be a clear borderline between both sexes. The dichotomy appears so natural to most people that they show amazement when they hear that biologists do not take this sexual dichotomy for granted.

What makes sexual reproduction so unlikely in the eyes of biologists is the fact that there is also a variant that is a-sexual. Members of an asexual species split themselves in two halves (e.g. bacteria) or they grow buds (e.g. yeast), which then mature to do the same again themselves. In a costs and benefits analysis, asexual reproduction would probably beat its sexual variant: It is much faster, more productive, and less risky. In this "rat race," sexual reproduction would end up as the loser, since the costs seem to outweigh the benefits: First, a good mate has to be found, which may be very time consuming; then there may be courtship costs in order to deal with potential rivals; and the mating procedure may create a moment of extra vulnerability.

And yet, sexuality can be found all over nature. In Darwinian terms, it must have had some advantages in order to become so widespread. What could be its advantage then over asexual reproduction? There are a few possibilities. The main one is that combining genes from two parents may create new combinations with some innovative potential in evolution. Another possibility is that having two copies of a gene makes organisms less susceptible to harmful mutations in one of them; plus reparations of the mutated DNA code would be possible via the unaffected code.

Whatever the explanation is, there appears to be some advantage in reproducing the sexual way, for some organisms (e.g., plant lice, also known as aphids) are known to alternate between sexual and asexual reproduction (which is called cyclical parthenogenesis). Depending on the situation, they are capable of switching to the other version when such is more advantageous—and apparently it is at times, otherwise they would not make the switch. Sometimes, the benefits of sexuality do outweigh the costs; sometimes, it is the other way around. Under certain circumstances, an established product may be less risky on the evolutionary "market" than a brand-new, untested try-out. At times, however, it could be worthwhile to go for a new and potentially better "trial."

However, the previous paragraph may explain that sexual reproduction did have a chance in evolution, but it does not explain how sexuality came into existence. How could asexual reproduction ever evolve into its sexual variant? There are many hypotheses, but no real winner. I think one of the best conjectures is that certain cells developed some kind of "lock and key" mechanism to make sure they would only "unlock" cells of the same species, so no genetic material would be wasted. You may object that this can very well be done with one type of cells (instead of a separate male and female type). True, but then this single type had to have both the "lock" and the "key," so cells of the same species could recognize each other and then "open" each other for fertilization. Perhaps that's how things got started, but then mutations began to create errors in the "lock" and/or the "key" part. The effects would be harmless as long as one of the two still worked properly, keeping it possible for the mutated cell to either recognize other cells or to be recognized by other cells. Besides, this would prevent cells with the same error from merging, for that would also block them from compensating each other's error. It is possible that this process eventually split certain species into two parts—the cells with only "keys" (let's call them male cells) and the cells with only "locks" (let's call them female cells).

From then on, the distinctions may have become more and more pronounced. The male half of the cells developed an evolutionary strategy similar to what insects do—creating many tiny descendents, but spending no time on the progeny. The female half of the cells followed an evolutionary strategy similar to what nest animals do—investing much time and energy in the care of a small number of relatively large descendants. And that's what we still find today: Males produce small sperm cells but in huge quantities, whereas females produce large egg cells but in tiny quantities. Does this scenario solve the riddle of sexuality? At least, it seems a fair trial within the framework of Darwinism.

The third "enigmatic" step is the origin of *humanity*. The first question is as to *when* this happened. Science has only access to the material substrate underlying the immaterial human mind. Since science can only go by physical appearances, scientists usually take the following four criteria into consideration:

1. Being a two-footer (biped) seems to be a big advantage for other human features—the main advantage being that the neck muscles put less restraint on the expansion of the brain part. *Homo erectus* is literally the "biped man," but does this fact make him/her a human being? Most scientists think being a biped in itself is not enough to qualify, but this decision is rather arbitrary, and hardly a scientific one.
2. Having a relatively large brain capacity seems to be necessary for most activities of the human mind. *Homo sapiens* is indeed the "rational man" who clearly has a relatively large brain compared to other

mammals. Brain capacity has often been used as an important indicator for being human, but we have no idea where to draw the borderline. Besides, brain mass or quantity is not clearly correlated to brain quality, let alone mental activities. And the human mind is a very peculiar entity, besides, as we will discuss later.

3. The jaw part of the skull should be proportionally small so the smaller jaw muscles would free the brain part to expand. In general, we do see a tendency for the brain part to increase when the jaw part decreases. But we do not know which one is the cause and which one is the effect, not to mention how else they may be related or connected.
4. The capability of making tools qualifies as an essential characteristic of human beings. Some animals may be capable of using tools, but creating tools seems to come with a more complex (and perhaps larger) brain structure. *Homo habilis* is literally the "handy man," but should we consider him/her a human being? Who is to make such decisions? Perhaps tool making is a necessary but not sufficient condition for being an "intelligent designer."

Does the previous knowledge make it easier for us to locate the transition point between pre-human and human beings in evolution? It might help, but it does not really aid us in assigning the proper status to specific paleontological finds. Notice how the aforementioned species names all differ, suggesting there is reproductive isolation involved—but we have no way of knowing such a thing. Apparently, there is a lot of guess work going on, sometimes based on almost "anecdotal" evidence of a single fossil find, so we should not place too much confidence in what scientists tell us in this respect. I would say, the "When" question has absolutely no reliable scientific answer. Looking at skulls gives us no clue as to when our ancestors became endowed with rationality and morality; that part of the process seems to be rather gradual. The problem is that the human soul and mind—and they make for the core of humanity—just are beyond anatomy and DNA.

Then there is the question *where* did this happen. Based on skeletal finds, most paleontologists would point to Africa as our cradle. But when probing further, we get trapped into many nasty questions. Paleontological evidence is rather "anecdotal" and in general very inconclusive.

Fortunately, we do get some better evidence from DNA research, because it lets us trace back what the DNA of our (first human?) ancestors must have looked like. This is done by comparing the "neutral" DNA sections of two types of DNA: mitochondrial DNA (passed on by the mother) and Y-chromosomal DNA (passed on by the father).[4] All lineages are identified by one or more so-called DNA markers (rare mutations in non-coding DNA segments) that create a "DNA signature" or *haplotype* of their ancestry.

These lineages are like "ancestral clans" and form *haplogroups* with some specific rare mutations and then split into subgroups that have some additional markers (or rare mutations).

This kind of analysis allows us to construct a DNA family tree that has at its origin a "mitochondrial Eve" and a "Y-chromosomal Adam" representing our most recent common ancestors that have DNA markers like ours. Such a reconstruction tells us that these ancestors were living somewhere in Africa (which is consistent with paleontology). All other contemporaries of these "proto-humans" failed to produce a direct unbroken male or female line to present-day human populations.

Although this DNA approach confirms that humanity originated rather recently in Africa, and not at several separate locations as some scientists had claimed before, it also has many limitations because it is based on several questionable presuppositions.[5] So we could give science some credit here, but always with due reservations.

The third question regards *how* it happened. It is part of Darwinism to stress the gradualness of this process and the usual impact of factors such as mutations and natural selection. I think this is a fair claim as far as our genome, our DNA, our anatomy, and our physiology is concerned. Indeed, even our brains don't differ dramatically from the brains of chimpanzees. In addition, it is striking that our genome doesn't differ much from our closest evolutionary relatives. We share, for instance, nearly 100% of DNA located inside genes with chimpanzees.

However, our *genome* may be very similar, but not necessarily our *genotype*, which contains the blueprint of a genome plus the instructions for processing that blueprint. As we found out, small gene mutations can have large effects. Take, for instance, the FOXP2 gene on chromosome 7.[6] Not only do the human and chimp versions of FOXP2 look different but they function differently as well. The FOXP2 gene switches other genes on and off, thus driving target genes to behave differently in both species. It could very well be that such seemingly minor genetic changes had a major impact on the development of language. That's Darwinian gradualism at its best.

NOTES

1. Christian De Duve, *Blueprint for a Cell — The Nature and Origin of Life* (Burlington, NC: Neil Patterson Publishers, 1991), chapter 5.
2. This theory was strongly advanced by Lynn Margulis in a 1967 paper, "The Origin of Mitosing Eukaryotic Cells," *Journal Theoretical Biology*, 14 (3): 255–274. It has received much attention since.
3. See chapter 9.

4. Mitochondrial DNA is passed un-mixed from mothers to all her children, along the maternal line; Y-chromosomal DNA is passed un-mixed from fathers to all his sons, along the paternal line.

5. I am referring to presuppositions such as a certain, unchanging rate of mutation and a lack of natural selection in "neutral" DNA sections.

6. See Collins, *The Language of God*, 139–140.

Chapter Eighteen

No Thresholds? A Not-So-Good Legacy

There may be gradualism in the development of human anatomy and physiology, as we discussed in the previous chapter, but I have my strong doubts as to whether this explanation covers the core of humanity—rationality and morality.

Why do I think rationality and morality are not a matter of biology? What makes these two so uniquely human? The answer is very straightforward: They are part of the human intellect, not of human intelligence. They are a matter of the mind, not the body. Biology can deal with the body, but not with the mind. The mind does not occur on the physical map of the body, and is nowhere in the chain of bodily activities—and yet it is the "soul" of it all and pervades the entire body. Let me explain.

Obviously, body and mind—or brain and mind, for that matter—have a rather mystifying relationship. It is definitely not a mental relationship; on the other hand, it's not a physical relationship either. Thoughts may make you cry, but certainly not in the physical way that a cut onion makes you cry; and crying may make you think, but there is certainly not a physical causation here. We should never degrade the mind's activity to a physical link in a network of physical causes between neurons, muscles, etc. If thoughts were physical, then we should be able to capture them on brain scans, and doctors should be able to read your mind by just looking at your EEG. If thoughts were really the product of bodily actions, all thoughts would be equivalent, and we would have no way of telling true from false, or knowledge from error. Apparently, the mind is not one of the players on the neuronal scene, but instead it is the author as well as the producer behind this neuronal play.

Obviously, the mind does interfere with physical causes, but it does so in a rather "mysterious" way. The mental and physical aspects affect each other similar to the way mass and gravity affect each other; we just do not know the "mechanism" of their interaction; somehow, their interaction is part of the cosmic design of creation. Think of this: What is the difference between blinking and winking? Physically, they are in essence the same, but the act of blinking is more than a physical process—there is an intention behind it. We are able to steer the laws of nature behind blinking with the intention of winking. But as to how the mind and its intentions interact with the brain and its causes eludes us. We just lack understanding of what enables our understanding.

But no matter what or how, humans definitely are "self-moved movers," altering themselves and their surroundings through decisions they make and through values they hold in their mysterious minds. As the Nobel-laureate and physicist Arthur Compton (known for the Compton effect) used to say: "If the laws of physics ever should come to contradict my conviction that I can move my little finger at will, then all the laws of physics should be revised and reformulated."[1] Even in a world ruled by the law of cause and effect, there is still our own ability to be the cause of events, although not in a physical sense. Thoughts can and do move the world as we all know. But how can this possibly happen?

Let me try to make this issue more "tangible" by using an example from sports. When golfers hit their ball, or when baseball hitters strike the pitcher's ball, they apply the laws of physics—that is, they use a specific force at a certain angle with a specific impact, and so on. We have a cascade of physical causes and effects here. And yet, there is much more going on in this process—these players have a very specific intention in mind, which eludes and transcends the laws of science. Do they go *against* the laws of nature? Of course, they don't; but they do go *beyond* those laws. People who can't look beyond those physical laws and causes are completely missing out on what the game is all about.

All I can say at this point is that the mind must be more than the brain—in the same way as a projector must be more than the images it projects or a copy machine must be more than the copies it makes. The German philosopher Gottfried Leibniz once suggested to picture the brain so much enlarged that one could walk in it as in a mill.[2] Inside, we would only observe movements of several parts, but never anything like a thought. For this reason, he believed, thoughts must be different from physical and material movements and parts. Put differently, how could the physical world ever be studied by something purely physical? Studying what is physical requires someone supra-physical or trans-physical, which takes us into the meta-physical realm—in this case the realm of the mind.

True, in response to certain thoughts, some brain areas may appear more active on brain scans, but that doesn't necessarily mean that those areas *produce* the thought; all we can say is that they most likely *carry* the thought. In other words, thoughts are more than brain waves. We can read and interpret an EEG, but looking at it doesn't show us any thoughts.[3]

I could carry this argument even further. An X-ray or an EEG is just a worthless piece of "paper" until the human mind interprets what's on it. It requires a subject to study an object; it requires a mind to study the brain. Without the "I of the beholder," nothing would make sense in life. And that's where our thoughts come into the picture as mental, immaterial, non-physical, and spiritual entities.

It follows from this that the knowing subject must be more than the known object; in our knowing, we have got to be more than what we know; we are rational and moral beings—and that's what really sets us apart as human beings. In addition to our five material senses, we have two immaterial senses—the sense of true or false and the sense of right or wrong.

If it is true that the mind—with its rationality and morality—is not of a biological nature, it cannot be an issue of genetics either. Neither rationality nor morality is based on DNA. If rationality were ruled by genes, we could skip many years of learning the rules of logic, mathematics, and methodology. If morality were a matter of genes, why would we need an articulated moral rule to reinforce what by nature we would or would not desire to do anyway?

Nevertheless, many scientists have tried to base rationality and morality on genetics. Let me show how they attempted this as far as *morality* is concerned. Some Darwinists keep stressing that evolutionary biology can explain how humanity acquired its morality. Their magic wand is the theory of natural selection, of course. Take the example of incest. There is an almost universal human taboo on incest—phrased in a moral law that says "Intercourse with very close relatives is wrong and hence forbidden." Well, these biologists would point out that inbreeding between close relatives tends to bring out recessive lethal traits and other afflictions that lessen the offspring's reproductive success. Hence they argue that natural selection has been promoting a genetic basis for behavioral avoidance of intercourse with close relatives. And that's where they believe this moral value stems from.

Or think of moral laws such as "You shall not kill" (the fifth commandment) or "You shall not commit adultery" (the sixth commandment). Again they have made the case that humans must be monogamous "by nature" (just like animals such as swans), since that would give the offspring a better protection; that's why natural selection did promote monogamy—which would make for a biological basis of the sixth commandment. Others have made the claim that killing members of the same species would undermine the persistence of the species and was consequently selected against by the process of natural selection. They even consider the moral value of paternal care for children to be a product of natural selection; their reasoning is that fathers who don't feel an "instinctive" responsibility toward their underage children would reduce their offspring's reproductive success. What all these cases have in common is that morality is supposedly based on genes promoted by natural selection.

Again, if this approach were correct, we have to ask ourselves the question as to why we would need an articulated moral rule to reinforce what by nature we would or would not desire to do anyway. As it happens, there are only far too many people willing to break moral rules when they can get away with it. Are they really fighting their genes? Think again of the responsibility parents feel toward their underage children. Is this a natural, "instinctive" responsibility that was promoted by natural selection because it presumably improves the offspring's reproductive capacity? Apparently it's not, given the fact that far too many parents try to ignore their so-called "natural" responsibility. Or take monogamy: If we were monogamous "by nature," we wouldn't need moral laws to protect family life. In other words, we are rather dealing here with responsibility in a moral sense. Responsibility is not a genetic, but moral phenomenon.

And if that's not enough to defuse genetic claims on rationality and morality, I have to add another reason why such attempts fail. If rationality and morality were really based on genes, we would be in deep trouble. Let me show this for *rationality*.

I said it before, when Darwin discovered the law of natural selection, he must have been "more" than the theory he discovered. Like a projector must be more than the pictures it projects, the knowing subject, Darwin, must be more than the known object, the theory he discovered. If he were not, he would run into a serious problem of circularity. Even if the theory of natural selection in itself is not the product of natural selection, it still is a product of the human mind (Darwin's, to be precise). And where does the human mind come from? Hard-core Darwinists will tell you that the human mind is just a product of natural selection, for natural selection supposedly explains everything in life, including the human mind. But if that were really so, we would find ourselves trapped in a vicious circle.

Darwin himself was somehow aware of this thorny implication, as we found out earlier (good for him, I keep saying, but he didn't think it through!). It is worth noting, though, how exactly he worded, and thus obscured, his problem: One might wonder whether "the *mind* of man, which has, as I fully believe, been developed from a *mind* as low as that possessed by the lowest animal, [can] be trusted when it draws such grand conclusions" (italics are mine).[4] In both cases, he should have used the word *brain* instead of *mind*. An animal doesn't have a mind but a brain, and a human being has more than a brain—a mind, that is. The brain may have developed from the animal kingdom, but the mind is not a product of evolution. If it were, we definitely should be questioning the validity of our knowledge—which necessarily includes our scientific knowledge of the theory of evolution. Our intellect must be more than a matter of intelligence.

I would like to rephrase this thought in more revealing terms: If one doesn't trust the rationality of human beings, one is logically prevented from having confidence in one's own rational activities—with science being one of them. This version of Darwinism would run into serious trouble, by cutting off its reason for reasoning! Once I take natural selection to be the only power shaping me and my mind the way it shapes my DNA, I would have reason to doubt what my rational capacities are really worth. And evolutionary theory happens to be fully dependent on these very capacities—which fact gives it a rather shaky basis. How could we ever trust the outcome of mere natural selection when it comes to matters of truth? I myself wouldn't put any trust in it!

This trouble would be solved if the human mind were not a product of natural selection. The human *brain* may be a product of natural selection, but that doesn't mean the human *mind* is too. As a matter of fact, natural selection must assume the human mind, but it can neither create it nor explain it. The brain cannot study itself, only the mind can study the brain. So the mind must have another origin…

Then I have one more reason to reject the claim that rationality and morality are a biological issue, anchored in our genes. I think most of us would agree that rationality claims to be objective, and strives for universal truths. However, such a position does not fare well with the claim of some Darwinists that rationality itself is the product of natural selection as well. If rationality would be the product of evolution, favored by natural selection, our genes would only make us *believe*—a shaky belief—that our rational capabilities have an objective foundation, but in fact they wouldn't; their only foundation would lie in our genes. So we would end up with a collective illusion foisted on us by our genes. If this doctrine were true, rationality would be as fragile as the DNA material it is supposedly made of. We would have no *reason* to trust our own reasoning.

Modern society has been tyrannized by the despotism of relativism, which arrogantly asserts that we have been born "blind" and that truth is beyond our reach. That's when common sense is not so common anymore. Indeed, intolerance can take on many faces. Most people think absolutism is intolerant because it pretends to know "the" truth. Well, it doesn't "possess" the truth, but must listen obediently to the truth—and that's what scientists are supposed to do too. Instead, one could easily make the case that relativism is actually the intolerant one, because it conceitedly rejects any claim of truth as unacceptable.

Something similar holds for morality: It claims to be objective and makes for universal obligations and absolute values. Whatever is called "good" in moral terms tells us what we *ought* to do—no matter what, whether we like it or not, whether we feel it or not, whether we want it or not. In this case, we speak of real moral values. Values are absolute, because they tell us what we ought to do as human beings, irrespective of any other objective; hence they are often called intrinsic values. They make somehow for universal, objective, binding prescriptions; they are ends in themselves—not means to other ends. What you *ought* to achieve (which is a value) is not necessarily what you *want* to achieve (which is a motive or intention), and what you want to achieve is not always what you actually *do* achieve (which is an effect). Values have their own absolute status of what ought to be achieved, so we shouldn't confuse them with what we want to achieve or do achieve.

Because of its absolute values, morality cannot be based on anything that is relative and non-moral by nature; values are ends-in-themselves, so they can't be based on other ends, since that would undermine their absolute moral claims. If morality were really a matter of mere biological success, without any objective basis, we wouldn't be morally obligated, but would only *feel* obligated; our genetic make-up would only have us *believe* that our moral obligations rest on an objective foundation—a collective illusion, so to speak, foisted on us by our genes. Again, such a foundation is as fragile as the DNA material it is said to be made of. We would have no *right* to claim any moral rights.

Of course, the next question would be where rationality and morality do come from then, if they don't come with our genes or through the umbilical cord? Are they perhaps man-made? If they were, they would have no universal, absolute, and objective status. Yet, many claim that moral values are subject to change during the course of human history—determined by a majority vote, so to speak. However, I would warn you not to confuse moral *values* with moral *evaluations*. Moral evaluations are our personal feelings or discernments regarding moral values. Yet, some people think that, in making evaluations, we create values in accordance with these evaluations. So when evaluations change, the moral values and laws are said to change. If that were true, our moral values would be subject to various cultural and historical fluctuations; morality would just be a matter of emotions, personal preferences, cultural trends, and majority votes.

In response, I would emphasize that evaluations are merely a reflection of the way we discern moral values and react to them—but do not mistake our evaluations for values; whereas moral evaluations may be volatile and fluctuating, moral values and laws are eternal, objective, and absolute. Think of the following comparison: Our current understanding of mathematical and physical laws constantly needs revision until we reach a better understanding of those laws the way they really are. In the meantime, we assume there are absolute and universal laws of nature, although we may not yet have fully captured them in our current understanding and our existing evaluations. Something similar holds for moral laws.

Let me illustrate this point a little further. A few centuries ago, slavery was not evaluated as morally wrong, but nowadays it is by most people. Did our moral values change? No, they did not; but our evaluations certainly did. Only a few people in the past were able to discern the objective and universal value of personal freedom and human rights (versus slavery), whereas most of their contemporaries were blind for this value. How often were the best people who had the clearest discernment of moral values persecuted by the mass of blind people! And yet, the advancement of humanity did often depend on these very people having a sharper and better discernment of moral values. It's not the values that change in the course of human evolution but their evaluations—that is, our subjective attitudes toward these objective values. Anyone who does not see their evidence is blind. Just like there are color-blind people, there are also value-blind people.

I know what we are discussing here goes far beyond Darwinism—perhaps too far for many. But at least we should keep an open mind for this meta-physical terrain outside the Darwinian territory. Even Darwin himself touched on its boundaries, but didn't dare (or wish) to set foot into that territory; he actually shied away from it.

The famous atheist and philosopher Jean Paul Sartre once said that absolute or objective standards of right and wrong—and I would add, of true and false as well—require an eternal heaven that would make them objective and universal. Put in reverse, there must be an eternal heaven for our moral values to make them objective and universal. Without such standards, we could never tell right from wrong. The same holds for absolute or objective standards in our reasoning; they also require an eternal heaven. Without them, we could never tell true from false. Somehow we need standards "from beyond" to be rational and moral human beings. Indeed, if there is a Creator, we would not only have a rational lawgiver—who guarantees order, intelligibility, and predictability—but also a moral lawgiver—who guarantees decency, integrity, responsibility, justice, and human rights.

If you have ever been on jury duty, you know how much jurors depend on their human faculties of rationality and morality. Unlike animals, we are able to make rational and moral decisions—which fact also entails the possibility of irrational and immoral decisions. Take rationality and morality away from us, and we are back where we came from—animals.

Some Darwinists have tried hard to take them away from us—that's the bad part of Darwin's legacy—so they basically undermined their own enterprise, perhaps unknowingly. Without rationality, we have nothing to steer our knowledge, and without morality, we have nothing to guide our actions. Once morality no longer controls our actions, Darwinism can easily end up in what is called *eugenics*—the study and practice of selective breeding applied to humans, with the aim of actively improving the human gene pool. I would say such an outcome was bound to happen: If humans are merely animals, there is nothing that stops us from breeding them like we breed animals.

No wonder this idea of eugenics soon became a brutal movement which inflicted massive human rights violations on millions of people. The "interventions" advocated and practiced by eugenicists, extreme Darwinists, involved a wide range of "degenerates" or "unfits"—the poor, the blind, the mentally ill, entire "racial" groups such as Jews, Blacks, Roma ("Gypsies"); all of these were deemed "unfit" to live according to their despotic dogma called "survival of the fittest." And this, in turn, led to practices such as segregation, sterilization, genocide, preemptive abortions, euthanasia, designer babies, and in the extreme case of Nazi Germany, mass extermination.

Should we blame Darwin's legacy for this? Not necessarily so, but it may be one of its extreme offshoots. If Darwinism is kept within its own boundaries, it wouldn't lead to such excesses, but it does have the potential to go overboard in its claims. Darwinism may ignore the issues raised in this chapter, as they are not of a scientific nature—not quantifiable, not countable, and not calculable, but meta-physical. However, if some Darwinists claim that those issues are therefore fake or non-existent, I would say that's where Darwinism has gone awry. Let's go for the good parts of Darwin's legacy instead. Science shouldn't control morality, but morality should control science in the same way as rationality controls science.

NOTES

1. A. H. Compton, et al., *Man's Destiny in Eternity. A Book from a Symposium (The Garvin lectures)* (Boston, MA: Beacon Press, 1949).
2. Gottfried Leibniz, *Monadology* (1714), section 17
3. This is very different from pain and other sensations; those can be physically induced with electrodes, but such a thing is not possible for thoughts, since they are of a different nature.
4. See note 116.

Suggestions for Further Reading

Ayala, Francisco J. *Darwin 's Gift: to Science and Religion.* Washington, DC: Joseph Henry Press, 2007.

—*Darwin and Intelligent Design.* Minneapolis, MN: Fortress Press, 2006.

Collins, Francis S., *The Language of God.* New York: Free Press, an imprint of Simon & Schuster, 2006.

Darwin, Charles. *The Origin of Species* (6th Edition 1872). New York: Macmillan Publishing Co., Collier Books, 1962/1978.

—*The Autobiography of Charles Darwin*, ed. Nora Barlow. New York: W.W. Norton & Company, 1958/1969.

Gould , Stephen Jay. *Rocks of Ages: Science and Religion in the Fullness of Life* . New York: Random House, Ballantine Books, 1999/2002.

Kass, Leon R. *Toward a More Natural Science.* New York: Free Press, an imprint of Simon & Schuster, 1985.

Kitcher, Philip. *Abusing Science: The Case Against Creationism.* Cambridge, MA: MIT Press, 1982.

Mayr, Ernst. *The Growth of Biological Thought.* Cambridge, MA: Harvard University Press, 1982.

—*One Long Argument; Charles Darwin and the Genesis of Modern Evolutionary Thought.* Cambridge, MA: Harvard University Press, 1991.

Miller, Kenneth R. *Finding Darwin 's God: A Scientist's Search for Common Ground Between God and Evolution.* New York: HarperCollins, 2007.

—*Only a Theory: Evolution and the Battle for America's Soul*. New York: Penguin, 2008.

Ruse, Michael. *The Darwinian Revolution: Science Red in Tooth and Claw.* Chicago: University of Chicago Press, 1979.

—*Is Science Sexist?* Boston, MA: Reidel Publishing, 1981.

Sober, E. *The Nature of Selection: Evolutionary Theory in Philosophical Focus.* Cambridge, MA: MIT Press, 1984.

Verschuuren, Gerard. *Life Scientists: Their Convictions, Their Activities, and Their Values.* North Andover, MA: Genesis Publishing Company, 1995.

Index

About the Author

Gerard M. Verschuuren is a human geneticist who also earned a doctorate in the philosophy of science. He studied and worked at universities in Europe and the United States.

He has written numerous books, articles, and textbooks in Dutch; he was the lead author of a series of biology textbooks for colleges and universities in the Netherlands. He is also the author of *Investigating the Life Sciences* (in the series *Foundations & Philosophy of Science & Technology,* (1986), and *Life Scientists: Their Convictions, Their Activities, and Their Values* (1995), and his latest book *God and Evolution?* (2012). His next upcoming book is entitled *What Makes You Tick?—A New Paradigm of Neuroscience* (2012).

www.ingramcontent.com/pod-product-compliance
Lightning Source LLC
LaVergne TN
LVHW101449200926
848483LV00001B/96
9780739175200